When I first met Jesse in Vancouver back in 2006, I immediately felt a kindred spiritual connection with him and found him to be an authentic wild man for Jesus who carried something different than most of my other Christian friends. Jesse has a faith that can move mountains as well as an incredible passion to bring the love of Jesus to the nations. For many years, Jesse and I traveled to the far reaches of the earth together. We were a missionary team, along with our families and with all of the others who traveled the world with us. I can attest to the stories that Jesse tells as I was present first-hand during many of them. I applaud Jesse for the vulnerability that he displays throughout his book, "The Journey Unveiled," and I hope that you are encouraged by his tenacity and sheer will to overcome any obstacle that would keep him from fulfilling his purpose and destiny.

Christian Michael Jung
Long-time Friend and Fellow World Traveler

Romans 8 reminds me, ...*that neither death nor life, neither angels nor demons, neither the present nor the future, nor any powers, neither height nor depth, nor anything else in all creation, will be able to separate us from the love of God that is in Christ Jesus our Lord.* (38,39 NIV)

In The Journey Unveiled I am also reminded that when we are at the highest point in our life and at the lowest point of our lives, there is nothing the love of God cannot breakthrough.

Jesse shows us an honest, raw life where God would not give in. Even when the world around us would choose to separate and disassociate, our loving Father would not even consider letting go. I have known Jesse, Kayla, and the kids since they made that fateful journey from the Big Island of Hawaii back to the Pacific Northwest and we have had amazing victories as well as deep valleys. It's during these times where relationships are tested friendships are either dissolved or strengthened. I am happy to say even though it was very close to destruction, strength came out of those valleys, and we are still great friends.

While you may read this account with eyes ever widening, I believe if you just stick it out you will see victory and the blood of Jesus woven through each chapter. When you really take the time to invest you will not be disappointed. Kind of like each of our own journeys, when we choose relationship instead of loneliness, God uses that to restore us.

Journey Unveiled is a restoration story, in as much as there is sadness, there is a resulting joy that cannot be won without war. Believe me when I say, I do not use the word war out of context. Jesse and his family fought a war. A war of addiction, self-harm, and personal destruction. But… right when the enemy thinks he's won the battle, he realizes he's on the brink of losing the entire fight.

Well done Jesse for not ever giving up on the fact, you were created for MORE. I truly believe the world is yet to see what your family is capable of. I am excited to experience the other countries we get to visit together and even more excited to see the fullness of the restoration process as the Lord continues working within you.

Geoerl Niles
Coauthor of *Dreams Matter*
Pastor at Bethel Leaders Network
Lead Pastor at The Calling Church- Vancouver, WA

THE JOURNEY UNVEILED

~The Honest Artwork of an Abstract Life~

-A TRUE STORY-

Jesse Gellatly

ISBN: 978-0-578-34274-0 (paperback edition)

Cover photo- photographer credit: Rowin Sims

Authors website: www.JesseGellatly.com

**I dedicate this book to Zoe Joy Gellatly
My firstborn and beautiful daughter**

When I felt I had no one and nothing left to live for in this world,
you were the light in my heart that caused me to breathe another day
and reach for my tomorrows.

May God fulfill and surpass your dreams beyond your wildest
imagination.

Contents

Acknowledgements

I want to thank my beautiful wife, Kayla Gellatly. You are my Sugar Muffin. Thank you for loving me, forgiving me, trusting me, and supporting me. Cheers to one thousand more adventures together and infinity tomorrows.

Thank you Mom and Dad for never ceasing to pray for me, believe in me, and supporting me and my family in the journey of life.

Thank you Tanya for continuing to mother Zoe, Zion, and Zuriel when I got lost into the oblivion of burn out, deception and addiction.

Thank you Pastor Geoerl Niles for receiving me back into your church family when I gave you every reason to never open the doors for me again. Thank you for believing in me and offering counsel as I began the journey of writing my story.

~INTRODUCTION~

(IMPORTANT: Read first)

This was not an easy story for me to write. It is an honest depiction of my journey through life written chapter by chapter. Both the highs and the lows are fully revealed with nothing left to hide. The story portrays God's endless faithfulness despite my frequent failures. This book is written in journal format so that my honest feelings and emotions may be expressed from those moments of time. There are seasons that may spark adventure in the dreamer's heart and there are struggles that may be difficult to read about.

Throughout the story I was wrong many times and these errors became learning experiences for me. With this in mind, may the reader understand that how I felt and what I said in several chapters is not meant to be a teaching of solid truth. Often, my experiences may contradict your beliefs and leave you with questions or in disbelief. This is okay. I do claim, with absolute conviction, that the entire story is written accurately to document my adventures and lessons I have learned in life. I believe I see things differently than most people and have provided the testimony of my journey through my perception. Each one of us is different than everyone else at the core of who we are. When we embrace our true identity instead of trying to portray an image we think others will admire, then we are able to learn from others and become complete as one unit together.

If you are willing to embark on this quest to read The Journey Unveiled, be prepared for a wild ride.

The Journey Unveiled

CHAPTER 1

Searching for Purpose

Keep in mind. The beginning of this book is written by 13-year-old me. I was lost, depressed, and became suicidal. As the journey continues, my real emotions and perceptions are written from those current perspectives. Chapter one is a long, dark tunnel with a light flickering at the very end.

THE CHOICE
May 30th, 1998

My family recently moved here to Spokane and I feel like I'm supposed to be so happy and excited about it. We have a huge house, an indoor swimming pool, and so much land. I'm not happy about it at all. I feel extremely depressed and angry. I'm thirteen years old and I was forced to move and leave all my friends behind in Portland. This has caused me to become very resentful and bitter toward my parents. I have no friends here; I don't get along with my brothers, and my dad still has to spend most of his time commuting back and forth from Portland because he hasn't found a new job here yet. Now I'm stuck in the middle of nowhere; I'm homeschooled; I'm bored with no friends, and I'm screaming on the inside while helplessly in a new situation I can't control.

I've been reading my Bible and devotional book in my room. I feel peace and comfort as I read, but I hide and read them when no one is around because I feel shame that my brother will make fun of me for it. I'm faced with a constant pressure from him to do the things I know are wrong. My parents are very strict. They have so many rules and I feel like I'm not allowed to do anything I want to do. What are they trying to hide me from? As I stand here at the top of my driveway,

I feel a major choice is being laid before me. Either I can continue to pursue God and surrender my life to Him, or I can rebel and begin experiencing all the things in this world that I've been sheltered and held back from. The world seems dangerously enticing and exciting to me. I want to know why I'm not allowed to know about the "the world out there".

I can serve God when I get older. Right now, I need to know what I'm missing. I choose the world out there.

DEATH'S APPEAL
January 15th, 1999

Who am I? Why am I here? There must be something more to life than this. There has to be a reason I am alive. I need to know what my purpose is. I tried to be cool and I failed. I tried girlfriends and I only embarrassed myself. I tried to be a high school basketball star and that was an awkward mistake. I keep trying anything I can to fill this empty place inside of me but the more I take in, the emptier inside I become. I've become a lost soul. My life is meaningless. There is nothing more to life than this. I am so miserable that I don't want to live anymore. I hate everybody and everybody hates me. I want to disappear forever.

I've walked through this snowy forest for hours. It's well past midnight. I'm all alone and nobody cares. I managed to climb over that tall chain link fence and all the way to the top of this giant water tower. I know why I came up here. This is my time to finally escape. I will jump now. They will eventually find my body, but nobody will miss me. They will be glad I'm gone. Something deep inside me hopes they will miss me. I hope that committing suicide will cause others to feel this agonizing pain I'm feeling now. I don't have any other way to communicate it to them. I will end my pain this way and then others might realize how much hurt I was feeling; after I am gone…

It's 3am now and I'm still up here. It's so cold and I keep nearly sliding off the slippery, rounded top of this tower. I still can't jump.

Something is holding me back and I don't know what it is. I can feel an invisible arm holding me around my chest. I'm going to climb back down now. I'm so far away from home.

ETERNAL DARKNESS
February 23rd, 2000
(This journal entry is a summary of my last week written moment by moment to illustrate an experience I will never forget.)

Smoking weed is something I told myself I would never do. I finally gave in and tried it, and now I need to have it every day. It has become a pleasant escape from this miserable existence. I keep hearing about other drugs people are using. I'm a little scared, but what have I got to lose? Our home is so big and beautiful, but I don't want to be here because there's a world out there waiting for me.

Nothing is getting better; only worse. I try to stay high on anything and everything so that I don't have to feel. I took the ecstasy pills I was given and just scratched myself all night until I was bleeding. I took the LSD, watched a movie, then sat there staring at the wall all night. It's all unfulfilling and I still haven't found what I'm looking for. I've now inhaled every can of spray paint that my parents own. It makes me have some crazy hallucinations. I experienced the moon coming down to play a game of softball with me one time. I hide everything from my parents. I am careful to always wash the paint from around my mouth so that they don't notice. They've caught me a few times. They were smart and hid the remaining cans of spray paint. I searched the house until I found the last two cans and inhaled both in one sitting. I was searching for a new experience and then I had one and hated it. I became the can of paint sitting in the bark dust. I was first terrified that I had no arms or legs. I then began panicking that my family would find me there in the morning and throw me into the garbage, not knowing that I was the can of paint. No one would know what happened to me. Suddenly it made sense to me why I was warned to not huff spray paint. After that, I became aware that I was in my body again and I never huffed another can.

People have described their psychedelic experiences to me after eating hallucinogenic mushrooms which made me eager to try them. I brought home a quarter ounce from my friend at church and I'm ready to have my own experience. My friend Viper came over and we are indulging together. They do not taste good, but I'm not eating them for the taste.

I still don't feel any effects, but this movie is funny anyway. WHOA! Everything in my room is coming alive, swirling and pulsating! Why is Viper turning all the lights off? Why can't the TV be on anymore? Where are all these creatures coming from? Who am I? What am I? What is reality? Death is on me now. I made the wrong choice. I can never go back.

Sadly, I am still alive. I don't want to be. I'm a worthless soul. I have no place to go. Nothing in life is real to me. Happy life is a fantasy. I feel I'm lost, but I know this place. What am I going to tell God when I see Him face to face? I cannot stay in this world anymore. I am going now. I believe in heaven and hell. In heaven, I think a bunch of church people just stand around and sing hymns forever and ever. Church is so boring, and I don't like hymns. I do not want to go to heaven. All my friends will be in hell and it will probably be fun forever. Whatever it is like, it is most definitely better than this dreadful, meaningless existence on earth. Bye world!

What happened? Everything is darkness. My body is gone but I am here. I can hear myself, but I can't feel anything. There is nothing. I can hear myself, but no one can hear me. There is no one to hear me. There is only me, and nothing. I succeeded. This is forever. I am sinking… and sinking… and sinking deeper… into nothing, forever. I can never go back. I will never see my family again. I made the wrong choice and I have no more chances. I am completely alone; forever. This is the never-ending blackness of darkness. I MADE THE WRONG CHOICE! I CAN NEVER GO BACK! I MADE THE WRONG CHOICE!

GOD! CAN YOU HEAR ME! PLEASE CAN I HAVE ANOTHER

CHANCE! I DIDN'T KNOW THAT IT WAS GOING TO BE LIKE THIS! I AM SORRY!

I 'm back in my body again. I don't know how, but God must have heard me even in death. I didn't experience eternal darkness for ten minutes, one hour, one week or even for a thousand years. I was there forever and then God brought me back to where I was when I first took my own life. That is not a concept that anybody can grasp with a finite mind. It is a reality that can only be comprehended by someone who has experienced it. I have. Every soul has an eternal existence. Eternal life or eternal death are the choices, and the choice is yours. I chose eternal death and God brought me back for a retry. If the only reason that He chose to do that is so I could advise you not to choose death, then it would be worth it. Eternal death is more terrifyingly agonizing than could ever be expressed with human words or imagined by the most brilliant minds.

POSSESSED

October 7th, 2001

I'm seventeen now. I am alive again and still miserable for every second of every day. I don't want to live, but I can't die… again. The only escape that I can find is to stay high… high on anything and high on everything. I eat psychedelic mushrooms often now and I always have bad trips. Every time I use them, I see the pit of hell open up in the most real sense imaginable. I stare at the millions of screaming souls, flailing about in the flames, screaming for mercy. Horrific creatures are grabbing at my arms and legs, trying to drag me in. Despite this, I still can't stop buying and eating mushrooms. I love darkness. I like using this broken glass to carve my arms up with pentagrams and anarchy signs. I worship Satan now. He controls me. I've been studying what I can about it on the internet when no one's watching. I believe that if I let these demons come inside of me, it will give me more power. What have I got to lose? Nothing.

I lifted my hands and prayed to darkness, repeating the phrase "demons come in" three times.

I'm okay. I have my friends (evil spirits) with me all the time. They are always talking to me so that I'm not alone. They guide me in everything that I do. I always listen to them and no one else can see them. They've sketched their own faces onto these three portraits through this pen in my hands. I can never normally draw this good. Everybody hates me. I hate myself and everybody. I cannot be around my family anymore. I have found the perfect and only solution to deal with this inescapable, miserable existence. I am going to perma-fry my brain on hard drugs now. This way, whether I have drugs or not, I never have to return to reality again. My mom and dad have finally let me leave home for good and are not trying to get me to come back. I'd rather just live on the street where I'm free to get high and stay high.

BORN AGAIN

August 18th, 2002

(As in a previous entry, much of this segment is written in present tense, which helps me best recount and describe my experience.)

Somehow I've managed to go from being a homeless street kid, to having a job, home, and car by 18 years old. I've made it this far and have been high on hard drugs every day, along the way. I've succeeded in my goal of a perma-fried brain. My invisible friends still always support me... I am lost.

HELP ME! I did too much, what's happening? My heart's about to blow and I have no control. I am overdosing again... and again, and again. I keep two beer kegs in the trunk of my car, so there ends up being a keg party wherever I go. I frequently find myself having combined ecstasy, cocaine, and mushrooms by the end of the night. How many times will I continue doing this, then calling out for God to save me, until He finally allows me to die and return to eternal darkness? I cannot go back there. This could be my last chance.

I DON'T WANT TO GO BACK THERE!

I'm not sure how I got back to my house after such a crazy night. I'm here alone now and it's 3am. There are sirens going off everywhere and I think it's the end of the world. I think the rapture must have just happened, and I've been left behind.

Suddenly, I hear a new voice speak to me. It is different than the others. There is peace and safety in this voice. When I listen to the other voices that have become so familiar in my head, my body feels tense and I gnaw my jaw. As I listen to this voice, it is leading me away from the other voices. I follow and obey the voice of peace as it leads me. I threw away all my drugs and paraphernalia. I turned my television onto worship music. I found a Bible in my attic and opened it up. For the first time in my life, I see and understand what it's saying. I understand light and darkness, death and life, good and evil, at a level I never saw before. My eyes opened and jaw dropped as I read the book of Romans. This voice continued leading me and had me take down the three portraits I sketched of the demons I had invited to live inside me. As I destroyed each of these pictures, my head would turn, and a large spit loogy would come flying from my throat, out of my mouth, and out my bedroom window. This happened without me intentionally doing it. In this way, these demonic spirits left me and will never return.

It's now 4:30am. The dawn of a new day is shooting sun beams up from the beautiful horizon. The good voice, which is the Voice of God, spoke to me again and said, "look down". I looked down on saw confusion, darkness, and filth. He spoke again and said, "look up". When I looked up into the beautiful sunrise, the voice of my heavenly Father spoke to me and said, "All this time you were searching for your purpose in this world, when all you had to do was look up at the One who created the world and in Me was your purpose."

I was just reborn. I have a purpose!

I cried for help from the place of the dead, and You heard my voice. (Jonah 2:2)

When I chose death, You stood in my way. When I forsook You to hide in the mud and eat garbage, You sought me out relentlessly. You never gave up on me and still never will. Your love has overcome my rebellion. Your passion snatched me from the coils of death and made me a king instead. Now I am Yours forever. You are my life, my breath.

Death wrapped its ropes around me.

The terrors of the grave overtook me.

I saw only trouble and sorrow.

The I called on the name of the Lord:

"Please, Lord, save me!"

(Psalm 116:3-4)

CHAPTER 2

A New Beginning

Today is a new day. Who I was before has since passed away; deceased and buried with yesterday. I am seeing everything through new eyes. The earth is alive and there is beauty wherever I look. I'm different now. I met God. He revealed Himself to me and I feel cleaner and freer than I've ever been. I don't know what to do next, so I must continue to follow Him.

I've been through tough times, rough times,

craziness of all kinds

Made some bad choices, listened to wrong voices,

Sought after deadly poisons

Which took me on trips to strange places

Where I've seen strange faces

Images floating in the air

That weren't really there

And I didn't really care

Caught in a dead stare

Drifting away to who knows where

Nothing was real to me because I hated reality

Didn't care if I became the next fatality

Why?

Because I didn't see a purpose in life

Searched, and just became more empty inside

Couldn't hide, though I tried

Who will show me if there is another side?

To this insane ride-- of life

So I hit rock bottom; nowhere else to turn

But a glimpse of hope appeared

and my heart began to burn

God revealed Himself to me

So that I could see

That life could be

Worth living

For to Him, it was worth giving

His only Son to die

To give eternal life for you and I.

STRIPPED BARE

August 28th, 2002

Since that day, I've completely changed. The curse words which had accompanied every sentence from my mouth completely vanished from my vocabulary. I continued buying drugs and cigarettes out of habit, only to realize I had no more desire for them and would give them all away to my friends. I started getting confused attempting to continue following the voice of the Lord after He led me out of darkness. I was trying to do the next right thing but unsure what it was. I think I will have to leave all my friends behind if I want to live out this new life. I cannot do this. I love my friends too much.

I've started falling back into the same things I used to do, only it's not the same. I am not happy here. Each time I use drugs again, I end up overdosing and facing death. It keeps getting worse and I don't enjoy getting high anymore. I do love my friends but if I don't let them go, I will never get out of this lifestyle.

Over the past few days, I lost my job, then my car, and then my house. I am homeless again. I just pawned my skateboard for $3 so that I could by a McDonalds cheeseburger. It was my only possession I had left, but I was tired of eating wild mushrooms out of the grass in the local park. I'm starving and that is all I could find to eat. Now I feel like I just ate my skateboard and it tasted so good, but I'm still hungry. I really want to completely stop using drugs now, and my brother David offered to get an apartment with me if I continue doing good.

FULL SURRENDER

September 27th, 2002

I enjoyed having a home again and living with my brother. It was going well at first, but that's over now. I'm in the hospital and I just completely surrendered everything in my life to God. I started using drugs again in my apartment and when I use, I cannot use just a little. I smashed a very large ecstasy pill on my glass table and used my card to cut it into a big capital E. A couple of my friends were with me and I think they had the impression we were all going to share it. I snorted the entire thing myself and inhaled that drug into my head. Shortly after that, I was taken to the hospital because I could feel my chest about to explode. As I squirmed around in this bed, I told the doctor I was dying. At that moment, in this bed, I told God I would go anywhere He wanted me to go and do anything He wanted me to do if He would give me one last, last, last chance to live again. He told me to call my mom, and so I did. She reminded me about a Christian discipleship program in Los Angeles and I committed to going. I am flying out in two days. God had to strip me of everything, including my own life, repeatedly, to bring me to this place so that I can now emptyhandedly receive His grace. This is the kindness and mercy of God. I chose death, but He won't let me have it. He is giving me life instead.

DISCIPLE

September 30, 2003

I completed my one-year commitment to this discipleship ministry. I want to stay. I'm afraid that if I leave, I will backslide, and then I will die again and die quickly. This fear is driving me to start a second-year commitment. I have a purpose and destiny. I cannot take any wrong steps or I may veer off course and miss my destiny. I cannot leave unless my spiritual leader tells me that God told him it was my time to go, otherwise I will be going outside of God's will for my life. This is what I'm being taught here anyway. My days are very structured from before sunrise until bed. There is no such thing as "free time" here. But that's okay because I'm told that I'm not capable of doing anything good myself. The only cost for me to live here is to obey all the rules. If I don't follow them well or if I talk rude to an overseer I will be severely disciplined, so I must be very careful all the time. My mind was so twisted when I got here that I continue to memorize scripture always so that my mind will be renewed. I've now memorized the entire books of James, 1Peter, 2Peter, 1John, 2John, 3John, Jude, 2Timothy, Ephesians, and why stop now? My brain was perma-fried, but now it's restored and the Word of God is alive inside me.

STEPPING OUT

September 30th, 2005

By my third year with the discipleship ministry, I completed the entire New and Old Testament surveys and became one of the main leaders. My third-year curriculum was to read through the entire Bible. We've taken multiple extended-outreach trips across the USA. We traveled in our large busses and set up a stage in the inner cities of America. We performed breakdancing and choreographed productions and shared the Gospel. I often spoke on stage and shared my testimony. I have helped lead the teen's division and adult division of our ministry. I've been the cook, the driver, the tool-man, the office worker and one of our private school's teachers. I have learned so much! I didn't leave after my second year because I was still afraid of backsliding. I can never go back to what once was. I have a purpose and destiny now and I can't allow anything to take me off course. I love the director of this ministry. He's taken me under his wing and I've learned so much from him.

My third year is now completed. It hasn't been an easy journey but I believe it was worth it. My mom had sent me information about a Bible college near where they live. I got very excited while reading about it, but I was also very worried the director would not approve. My biggest fear was that if I told people I was leaving, everyone would think I'm leaving the will of God and that I'm in sin. I always pray and I started praying a lot about this. Deciding to approach the director about it was extremely hard. Most people sneak out at night if they want to leave because they fear his and other people's reactions. What we've been taught here is that he must come to us first and tell us that God spoke to him to bless us out of the ministry, otherwise we are running away.

When I approached him with my decision, he did not fight me about it. Rather, he seemed supportive and is giving me his blessing to go back to my family and attend Bible College. The other disciples who've been here with me for years are surprised by this. A lot has been dependent on me here and I feel like I'm hurting others by moving on. Nevertheless, if this is God, then I will step forth into a new chapter and trust Him. The director drove me to the airport and thanked me for how I have blessed his ministry. Now I'm waiting to board my plane and feeling pretty nervous.

He's turning the page again

I don't know how this unfolds, but I'm at rest in Him

He's never not been good

When I try to dictate my own course, I drive into the ditch

He takes me into Him

Where I see He's always been all I need in every time

I'm groaning deep within

He's stirring my spirit to desire more of Him

I'm thankful for the pain that drives me to my knees

crying out to Him

Because I belong to Him.

Therefore, anyone is in Christ, the new creation has come. The old has gone, the new is here! (2Corinthians 5:17)

CHAPTER 3

The Seasons Change

A NEW ADVENTURE

December 10th, 2005

My parents picked me up from the airport and showed me where my new bedroom is. I feel both nervous and excited. For the past three years I've lived in a very structured and protected environment. Before that, I was on drugs and out of control. I think I'm somewhat fearful of falling back into that pit. I fear personal freedom. I have experienced the depth of hell and now I'm saved. More than anything else, I want everybody to experience this same grace and mercy that I've been given. I want everyone to know Jesus and I need to tell them about Him.

I got enrolled into Bible College right away and started work for a temp agency. I don't have a car, so I take the bus and ride my skateboard everywhere. I've looked up all the churches in town and attend services nearly every day of the week. I love it! Each denomination carries strength in a different aspect of the nature of God. I'm thankful to have the freedom to venture out and explore the Body of Christ, because I hadn't had this opportunity before.

I'm twenty-one years old and more alive in Christ than ever. While staying busy, I met a new friend. I was riding the bus to work and saw a young man on the street corner dancing around with a Jesus sign. I had to get off at that bus stop and go talk to him. His name is Jason. When I asked him if he had another sign, he went to his trunk and grabbed a couple out for me. Soon after, there were a few of us out there holding signs together. Many people honked, some gave us the middle finger, and others stopped to ask for prayer. Jason had quit his job to do this instead. Now I meet him on the street every day after work and we hold signs together. I've made a handful of new signs too. My favorite says, *"Jesus is Coming Soon. Are you ready?"* Our group continues to grow. Now we are doing this in Vancouver and Portland. We take propane grills and feed the hungry, pass out Bibles and gospel tracts, and often have up to fifteen people in our group scattered across a couple blocks.

THE SCHOOL REBEL

May 2ND, 2006

My school semester is halfway over. I'm still doing good and staying out of trouble. I take public transport from my parent's house in Vancouver to my school in Portland and back every day. That's nearly four hours a day I spend on public transport. I'm known on the streets as "The Sign Guy", because I take my Jesus signs with me everywhere and I hold them up whenever I can. God has used this to provide opportunities for me to lead many people to Him while on the bus and the street. People at my school think I'm crazy because a lot of them believe that "friendship evangelism" is the only correct way to lead someone to Jesus. I'm kind of a rebel at school because I can be very outspoken in class against certain extreme conservative doctrines that downplay the power of God. I love being a student here regardless. I've become friends with the other front-row extremists and passionate ones after the heart of God.

One of my new friends just invited me to go to Hawaii with him this summer to help his friends plant a church among the hippie community. I am going. They're going to help me with my plane ticket because I don't have any money. This seems like a big step and change of plans for me, but hey, I'm always up for adventure and it's Hawaii. I prayed about it a lot first, as I do everything now, and I believe that I'm supposed to go during summer break and help the team out.

ISLAND LIFE

September 28th, 2006

Hawaii is beautiful and warm! It was my second time on a plane and I loved flying over the ocean. I only had $40 and my backpack when we arrived. We went to a small village town to ask for directions to the Hippie farm where we would set up our camp to stay. I noticed a Christian symbol on the sign for the Inn at the village, so I decided we should ask there first. We met the owner of the place in the back and after talking to her, I had a job, a rental unit to live in, and a new friend who loved Jesus. I'm now the maintenance man for the Inn, the restaurant, and other rental units in town. I've learned that when I step out in faith, God meets me there with provisions already prepared. The church planting team is a small group of young people with kindred hearts. The pastor greeted us with beers on the table. I was kind of nervous because I hadn't tasted a beer in years, but I drank it with them cordially. My new friends have been showing me all around the island. I love it so much! It's beautiful every day. We go to Hippie gatherings and raves, watch the fire dancers, and share Jesus with people. When I share the gospel with people while they're frying on mushrooms, they often seem way more open and receptive. Our group is growing.

My older brother, David, came to live with me in Hawaii. I love him being here, but my departure ticket date is set for August 10th. I'm supposed to go back to school for Fall semester. David does not need to go back to Portland yet and I don't either, except for school. More importantly than this, I heard God tell me to cancel my ticket. The church-plant is going ok. This corner of Hawaii is known to be one of the darkest and most difficult places in the state. Many church plants have started and failed here. Since I surrendered my life to Christ at age 18, I've done well at not pursuing any relationships with girls. I keep protected walls up because I was taught that it's important to not marry the wrong person based on feelings. So, what is crazy is that I now have plans to get married. It's silly how the lady I'm working for already introduced me to people as her son-in-law before I ever met her daughter, Tanya. She returned home from her ministry school a few months after I arrived here.

I feel bad that I don't spend much time with David anymore now that Tanya is here. We were both very reserved toward each other at first. It was clear to us that her mom had an agenda to put us together since she first got to know me. Tanya also started working for her mom, so we often get put on the same tasks. We didn't talk much until we started talking about Jesus and world missions. She has a dream to go to Africa and be a missionary with Heidi Baker. We naturally started acting like a couple after spending so much time together, and when I said, "when we go to Africa together next year", she was caught very off guard. I hadn't been part of her plan before. This evolved to, "If we are going to Africa together next year then we need to be married first, so does that mean we're engaged?" That's how I proposed to Tanya. I didn't mean to, or plan to, or think much about it. It just happened. I spoke with her father to request his blessing and his only terms were that we have nine months of premarital counseling first. There is hardly that much time before the missions training school will start in Mozambique, so we need to get moving. After talking about it together, we decided it will be better to go to Washington where we will have better assets for work and preparations over the next months. We're getting ready to leave the island and my parents have offered us separate bedrooms in their home during our engagement, while we go through premarital counseling.

AFRICA PREPARATION
June 2nd, 2007

Tanya and I are working three jobs each to save money for Africa. It's going to cost at least $6000 to pay for tuition and plane tickets. One lady has committed $125 a month to support us when we go. We had weekly pre-marital counseling with my friend Christian and his wife. We usually just worshipped for our entire counseling sessions. He started skate-church here in Vancouver and I love being on his staff team. These skateboarder kids are praying in tongues and doing miracles now. We have our passports, visas, and our flights booked. Our wedding was on May 26th, exactly one week before we leave for Africa. Christian officiated the wedding on the beach in Oregon.

I'm married now. Life always feels like a crazy thrill ride when I let my Papa God drive. The ministry we are going to be with has a rule that to become a missionary with them, you must attend the school, return home, then reapply for long-term. We don't want to come back because we already know we want to stay long-term and coming home would be a waste of a trip. Nevertheless, we have bought return-tickets to comply with their rules and are leaving it in God's hands.

Here I go again!

Leaping off the edge of this cliff, not knowing where I'll end

But in Your arms again

I didn't know I had these wings until I jumped to what could have been my end

And now I soar to higher height's than I have ever been

CHAPTER 4

The Other Side of the World

PEMBA, MOZAMBIQUE

August 18ᵗʰ, 2007

I enjoyed getting my first set of clothes splashed on with lots of poop after arriving. We were driven from the airport to the children's center and went straight to the tent for class. Bill and Beni Johnson were there teaching and prayed over us to receive impartation. Straight after class I was given a shovel and sent with three men to fill in the latrines with dirt. A latrine is a large man-made pit, securely covered with a small opening overtop. This is where everyone uses the bathroom by standing or squatting over the hole. When the pit is full of sewage, it has to be uncovered and filled in. This was my first assignment in Africa and with every shovel of dirt I threw in, the human excrement splashed out. It was super fun!

I stayed at Heidi Baker's main children's center, called Joy Village, for the first three months. It is directly across from a beautiful beach on the Indian Ocean. We were made the house parents in our tiny cabin with eight other women. There were about 150 international mission's students, many Mozambican children who were adopted by Heidi, many Mozambican Bible school students, and a handful of missionaries all living together on the compound. We all eat together in the cafeteria and have bread with tea for breakfast, rice and beans for lunch, and rice with cabbage or fish for dinner every day. It's so hot here in Africa. We have fans and limited electricity but often the power shuts off and the heat becomes unbearable under our mosquito nets. I have to continually splash water over myself in those times. Water shortage is another frequent issue. It often shuts off for days and we have to walk down the hill and draw water from the well then carry it in buckets back to our house. This helps me better understand how Mozambicans live every day. The women carry water, often for miles from the nearest water source to their homes every day so that they can cook, clean, and bathe their children. Several of the international students have gotten malaria from mosquito bites. Even people who were on anti-malaria medication still got malaria amongst several other illnesses we are constantly faced with.

People are coming from all over the world to this place for a reason. It was poorest country in the world and has been stricken by war, floods, and famine. Mozambique was not a Christian nation, but something is happening. The Gospel is being preached and the love of Jesus is flooding the land. People are frequently being raised from the dead. Blind eyes are being opened, the deaf are hearing, thousands of churches are sprouting up, and miracles beyond imagination are a regular occurrence. True revival is ablaze and spreading like wildfire from this place. When we visit the remote villages, we bring our tents and stay the weekend. The entire village gathers to watch the Jesus film on a big screen, translated into their local dialect. Many run forward to receive Jesus into their hearts after the film. We pray for the sick and there are always multiple miraculous healings. Witch doctors, village chiefs and Muslim Sheikhs surrender their lives to Christ. Often, whole villages repent of witchcraft after seeing the power of God manifest in a greater way. The fire of God is so heavy in our class times that most people are usually shaking, crying, laughing, or laying in piles on the dirt for hours. Some students fall into heavenly visitations and encounters that last for days. Others will carry them to and from class so that they are not just in their hot little cabin the whole time. It appears weird to most, but it is God. There are usually students who are slain in the Spirit laying in the dirt around the housing compound. This is normal here. God is wrecking all of us in the most bizarre and life-changing way imaginable.

We just spoke to the long-term missions committee and they gave us approval to not return home after graduation, but to stay on as long-term missionaries. A missionary family from the neighboring province shared at the school a couple weeks ago. They shared their hearts, vision and about the work God has given them to do with the Yao people of Niassa. Our spirits immediately felt connected with the Wilcox family. We spoke to them and they gave us their blessing to come and live with them at the new base they're starting. It's now worked out that after graduation we will go with the extended outreach team there and we will remain when they return after a couple weeks. We will be the first long-term missionaries to join them in that province. The Yao people are the most unreached group in the country, which is what causes me to be most excited about going there. I want to share Jesus with those who have never heard of him before and now I am being given the opportunity to do that with an entire people group.

REACHING THE UNREACHED

January 27th, 2008

It took a couple full days of driving over bumpy red dirt roads to get here. We traveled in the back of two large flatbed trucks. The other truck had a covering and ours did not. Rain poured down for most of the journey. I could see it beating against the covered truck behind us, pouring down to the sides of us, and drenching the road in front of us. When we finally stopped, the people from the covered truck came to check on us and were very worried and sorry that we didn't have a cover. They were surprised because we were all perfectly dry and they were soaked. It never rained on us. The Wilcox family are awesome. They are from Australia but served in Spain many years before coming to Mozambique. We will be staying in their little house with them. I went to the market and got a bed woven from reeds with a mattress made of grass stuffed into potato sacks. It's nice because there's no running water or electricity here, so I can get exercise by pulling our water with a rope and bucket every morning, and we have candle-lit dinners every night. I was given the position of management over the construction of the children's center here. I previously had construction experience but I have now learned how to build with cement and cinder blocks. I always love learning new things!

I'm adjusting to the way of life here in Niassa. We live down a red dirt road just past the airport. We are helping the Wilcox family finish building their house and are building the girl's home for the orphans they will be taking in. There's a lot of open land around us and we are overlooking a nearby village called Assumani. My primary responsibility is construction, but every day I pray over this village as my heart longs to go share the gospel with the unreached. There are missionaries from other organizations in town. They tell me about how hard it is to reach these people and how after many years they have little or no converts. They make it sound very difficult, but I believe my God is bigger and the fields are ripe for harvest. Victo is a young Mozambican believer who lives here and serves with us. He has big eyes and is always happy. I am teaching him English while he teaches me Portuguese and Yao. I learn new words every day! I asked Mr. Wilcox if we were able to start visiting and ministering in Assumani and he gave me the okay.

We purchased two bicycles from the market and started visiting the village regularly. The people there were afraid of us at first. All the women and children ran and hid when I showed up. They were told stories about how white people would eat the children and they believed the stories because maybe that happened in the past. Now, whenever I enter the village, all the children and many adults come running from every direction shouting our names joyfully. I often walk through the village playing my guitar while crowds follow, singing and dancing with me. There's a man named Phinehas at the gateway of the village. He warmly welcomed us into his home and family. He shared how he'd seen a heavenly vision where two angels came to him with a book. He asked them what was written in the book and they responded that someone will be sent to show him. Phinehas told me with confidence, "You are the ones the angels told me were coming"! He was eager for us teach him from the Bible. He and his entire household were saved. This set him apart from the rest of the village and brought persecution because of a difference of faith. When others in the village decide to follow Jesus, they often come and live at Phinehas's house because they become ostracized from their families.

Other missionaries are gradually coming to join our family in Niassa. We finished building the girl's home and Mr. Wilcox promoted our position from construction to evangelism directors. He recognized the hand of God redirecting us because of the large number of people from the villages who are turning to Christ. We ride our bicycles for hours a day to Assumani and the neighboring villages. Each village is afraid of us at first, then loves us dearly after they realize we're not there to harm them. We purchased a piece of land from the secretary of Assumani and are preparing to build our first church here. Our numbers are constantly growing and the little church hut where we live is not sufficient to be a church for the village. There are hundreds more villages across this province that are full of people who've still never hear the Good News. I want so badly to tell them all because people are dying every day without having an opportunity to hear about God's love through Christ Jesus. We still only have our two bicycles, so I've been praying for motorbikes, as well as a flatbed truck and sound system to show the Jesus film. When I pray, God keeps telling me that if I'm faithful in the least, He will entrust me with much. So, I will just keep going as far as I can with this bicycle.

INCREASE AND MULTIPLICATION
October 15th, 2008

I was sick with Malaria for weeks. I usually tried to make it to the outside latrine to use the bathroom. We have a little toilet in the house that has to be flushed with a bucket of water. There's no door on our bathroom; only a sheet hanging in front of the toilet, so it can be embarrassing when living in a tiny house with two other families. There are fourteen of us in this small block house. Diarrhea and vomiting were constant. I had a very high fever and everything hurt everywhere so much. I fainted while coming back from the latrine and Tyren Haynes caught me just before I smashed my head on a post.

While recovering, I continually listened to sermons on my mp3 player and studied Portuguese. I can now speak fluent Portuguese and conversationally in Yao. Christian is planning to move his family from Vancouver, WA to come live with us in Mozambique next year. The elderly woman who lived in the only other house on this hill recently passed away. I'm going to renovate the house to be a Bible School for the new believers. Whenever we go to the villages, more of the Yao people believe in Jesus. Often, we just stand in the road eating mango's and one by one someone stops to greet us, hears the gospel, believes, and commits their life to the Lord. We are the first white people many of the women and children in the villages have ever seen. I'm getting used to being called "Whitey" and men always asking Tanya to marry them.

Praise God we were able to purchase our first motorbike! I had never ridden one before, so I accidently did a wheelie as I pulled away from the market. It made the crowd scatter and laugh. I later gave the motorbike to our Malawian cook, Moses. We didn't have a motorbike anymore, but Moses was able to ride to and from the market quickly to get all the food we needed. Now I have eight motorbikes and we put 2-5 people on each bike so that we can go to further villages. I don't know how; just God's faithfulness I guess. I am getting a truck and sound system next. What is too big for God?

It only took five days to build our first church in Assumani. I paid $200 for all the materials and the men helped build it. It's made from tree posts, bamboo, black ties, plastic, and grass. It's strong and can hold about 200 people. Sometimes it has that many people in it on Sunday morning. I use our sound system and we have dance parties to Swahili worship music, share testimonies, dramas, and preach the word. We baptize people in murky fishponds and deep mud puddles. Oh, and we have a nice sound system now! I called the missionary who oversees the sound department in Pemba regarding questions about where to pursue one. "We just had a new complete sound system and generator donated to us!", she exclaimed. She arranged to have their pilot fly it over to us on the Iris plane a few days later.

God has been blessing us with provision in many ways recently. One of our biggest answered prayers is that we now have a 1-ton flatbed truck. The new believers in all our villages have been praying for months with us for it. We named it Chikulpidilo, which is a word in Yao that means faithful. We had $6000 donated, which was the exact amount of money we needed, and we received it on the day we needed it. We flew to the Southern tip of Mozambique to find Faithful. We prayed that it would be the first truck we look at, and then we looked at many trucks. At the end of the day, we went back and bought the very first truck we had looked at. Peter Wilcox received provision to purchase a pickup at the same time as us. It's a three-day drive to get home from Maputo. The night before our journey home, I woke up at midnight hearing banging on the truck. The truck was not close to me, so I walked outside the base to where it was parked. Three cats ran from beneath it when I approached. Every dog for miles in every direction was howling, and at the same time every rooster in every direction for miles was crowing. It was very strange, so I prayed over Faithful and went back to bed.

We left early the next morning and the journey was going well until 50 km the road, Faithful overheated. We returned to the Maputo children's center and discovered the radiator was clogged and full of junk. I removed and cleaned it. We set out again after I put it back together. 50 km down the road, it happened again. I took the thermostat out, let it cool, filled it with water and set out again. 50 km later, the same thing happened. I responded the same way, and this continued to happen every 50 km until we got to Vilankulos. There was a South African mechanic shop there and I asked them to inspect it. The office administrator came an hour later to give me the information. She was very sorry and explained how the motor head was cracked in many places. "But it's okay", she said, "We can have a new one shipped from Japan and it will be here in a couple days. It will only cost $3400". I only had enough money left for gas to get home. We had to leave Faithful at the shop and catch a ride back with Peter. We decided to fast for the next day after arriving home and pray for provision. As soon as we opened our mouths to pray that morning, the phone rang and we were notified of a donation being made to cover the entire cost. "I guess that means we can eat", I said. I notified the shop in Vilankulos and they called a week later to say we could come pick up the truck. I purchased flights for Peter and me to fly to Vilankulos so that he could help me drive it home, since it was still another 2-day drive away. When we arrived at the shop, the office lady was very sorry and said it was supposed to be done already but the plane it was on had a mechanical problem and was delayed. "It will be tomorrow", she said. Tomorrow came and the part was accidently put on the wrong plane and had to be sent back. The next day came and there was a storm so the plane could not take off. This continued for two weeks while Peter and I slept in the shop and waited. Peter needed to get back to base. He had a family to take care of and a ministry center to run. The office lady skipped into the room and happily declared that the part had successfully gotten on the plane and was in the air. The mechanic could have it installed within the day. We rejoiced and thanked the Lord. Thirty minutes lade she came back in the room sad again. She told me how it went to the wrong airport and had arrived at a different town five hours south of us. I decided to handle it myself and told Peter I'd be back. I jumped on the next minibus and took the extremely rough journey to Inhambane. I arrived at 9pm and learned that the shop where the part was sent to was across the bay in the town of XaiXai. I hurried to the bay dock and there was one boat left. They told me they were done for the day. I begged them and paid them extra to wait for me while I ran through the town, grabbed the new part, and ran back to the boat. The guard at the shop was waiting for me and gave me the package. It was big, awkward, and heavy, so I put it on my head and made a quick stride about a mile back to the dock. The town was mostly dead, except for a few late-night bystanders watching a giant white man with a big box on his head jog through their town at that late hour. I made it back to the shop the next morning and the mechanic went to work. Two hours later, the office administrator came to apologetically let us know the mechanic was careless and accidently broke the new head while putting it in. It was irreparable and they would have to order another new motor head from Japan. "But it's ok, we can wait there, and it should only take two days", she assured us. Peter and I hitchhiked back to Niassa. Hundreds of people were sad to see us arriving back, truck-less again. Tanya was going through her first miscarriage at the same time we started hitchhiking back to Niassa.

Three months later, the shop called to inform us that the new part had arrived, had been installed, test driven, and that everything was working great. Tanya and I flew back down to Vilankulos and drove Faithful home.

RADICAL RAGING LOVE flows like liquid fire, filling the darkness with hope and light, melting the ice, breaking the chains and calling forth a new generation to extreme living. Dying to self and the world, living new energy of passion bursting forth to love the ugly unto beauty and uncover the hidden treasures from the dust of the earth. His kingdom within reach, within us, wanting to be let out as we step out of comfort into the tension of the unseen and untouched being more real than the tangible and visible, until faith creates the would dreamed for- becoming the manifested physical reality

NEARLY STONED TO DEATH

February 11, 2009

The Gospel is spreading like wildfire across this formerly unreached province. I take Faithful, my little one-ton flatbed truck, and we travel village to village, showing the Jesus film, sharing the gospel, and praying for the sick. It's so nice to have Christian here. There is a major acceleration when we run together. People everywhere are turning to Jesus. We always have a team of our Mozambican disciples with us and they preach, share testimonies, and pray for people. We camp in the chief of each village's courtyard. Every chief usually asks me to breakdance first because word has spread everywhere that the tall white man can spin circles on his hands. I respect the chief's wishes, bust some moves, get dirty and sometimes bloody, then we set up our tents, stage, and sound system. We have been warned about a village called Micoco. People tell us not to go there because they are hostile and will harm us. Their chief invited us and we went there to show the Jesus film and stay three days.

Micoco is a very challenging village to get to. The way there is primarily just for bicycles and the road is very washed-out by the rains. We went prepared with shovels, a pick, and planks to cross the deep trenches. It took nearly three hours to get there from our house. One area of the road is so bad that Victo got out of the truck and slowly guided me over and around deep washed-out ditches. There were about 500 people present for the film the first night. One drunk man came forward at the end to respond to an altar call. The next day we went house to house, visiting and praying for people. There are three mosques in the village and in one area, the leader of the mosque was very angry and made us leave. We were asked to show the film again the second night. Soon after it started, my friend Luciano went to use the latrine at the chief's house behind us. He noticed a secret meeting taking place between the chief, secretary, and Muslim leaders. According to the leader who had chased us away early in the day, we have been burning down mosques in other villages and last night they saw us drive our truck to their cemetery and dig up a couple of the graves. This is a very sacred offense to them and there was testified evidence of bodies missing. The chief affirmed that he had seen a pick and shovel in my truck when we first arrived. They all agreed that they had to make an example of us and planned to drag us out of our tents after we went to bed that night and kill us. We discussed how we should respond and decided that we would pack up the sound system after the movie ended and leave that night, without giving any notice. A family member of the chief overheard us and ran to tell the village leaders who were still in their meeting place. They changed their plans quickly and decided to gather a mob and stone us before the movie finished. Thankfully, the chief's niece liked us and ran to let us know that we needed to leave immediately. I didn't want to run in fear, but suddenly a mob started forming and many people were scurrying to gather stones. I quickly said bye on the microphone and we tossed our speakers and generator into the back of the truck and everyone piled in. The chief begged me not to leave, as there would be bandits waiting to ambush us up the road. I drove fast back and forth around the large village, lost for a bit. It seemed like we were in the midst of chaos and fleeing for our lives. Tanya prayed, "Lord, give us wings". I left the village and pulled into our home about 20 minutes later. It seemed like I was waking up from a dream when we got home. We had never even come to or passed the most treacherous parts of the road on the way home. It took three hours in daytime to get there and only 20 minutes to get home driving late at night.

The nephew of the chief of Micoco came to my house today. He brought a message from the chief that they discovered the accusations made against us were lies and they "took care of" the person who was responsible. They are pleading for us to return, show the film again, and start a church there. So that is what we are going to do.

A light shines bright in the darkness

A new day dawns on the horizon

Brilliant colors radiate the magnificent majesty of Almighty

A rising sun sends forth rays of new hope for a day of unsuspected surprises

Negativity lurks its way into the unaware mind, sowing seeds of expectations for disappointments

Guard your castle and tend your garden so that the goblins of grumbling don't become your houseguests

Hope does not disappoint and is the joyful expectation of only good

Faith creates the fulfillment of your dreams

Fear is the creator of your fears

Place a notice of continual resistance to the seeds of fear, often disguised as worry, caution, or practical wisdom

Fear of the unknown hold's masses enslaved to the familiar

Faith in the All-Knowing possesses the keys to infinite storehouses of eternal treasure.

CHAPTER 5

Vison Expansion

Disappointment and Hope

March 1st, 2009

This is our first visit to America after living in Africa for nearly two years. Many churches have asked us to come speak at their services in Washington and Hawaii. Our calendar is now fully booked with ministry for these two months and it appears we won't have any time for rest.

I've heard about Bethel Church and listened to Pastor Bill Johnson's messages over and over with my iPod while in Mozambique. This is my first time being here in person at Bethel in Redding, CA. Tanya just went through her second miscarriage while flying over the Atlantic. We hopped on a train the day after we landed in Portland and made it down here for the Open Heavens conference. We went to the front of the church and asked Pastor Bill for prayer. He prophesied over us that Tanya would conceive within the year, there would be no more miscarriages, and that we did not need to worry.

Rain in Barren Places

Kenya, November 7th, 2009

I hit the ground running, blown by the winds of change. Upon return to Africa, we helped lead another missions training school in Pemba, returned to our home in Niassa, showed the Jesus film and preached the Gospel to many unreached villages across the province. Hundreds of Yao Muslims gave their lives to Jesus and were healed of various sicknesses and curses. I had been invited to Kenya by a local pastor named Vincent. His persistence led me to pray about it and after hearing from God, we brought Moses, our Malawian friend from Mozambique and flew into Nairobi, Kenya. We just witnessed the most incredible miracle! I am overwhelmed and need to write about it.

This land here has been in severe draught for two years. There are dead cattle along the roadsides. The people are starving and many are dying. I preach on 2 Chronicles 7:14 wherever we go.

If my people, who are called by my name, will humble themselves and pray and seek my face and turn from their wicked ways, then I will hear from heaven, and I will forgive their sin and will heal their land.

We spent some time in Mombasa but have mainly been with the Samburu people in the Rift Valley. They are warriors who resemble the Maasai. High in the mountains, a tribe gathered in a school building to see the visitors who had come to them. We were the visitors. I shared the Gospel and everyone eagerly responded to know Jesus. We prayed for the sick and people were being healed and filled with the Holy Spirit. Suddenly no one could hear each other anymore. It began to rain so heavily on the tin roof that everyone just started laughing and dancing. When we went outside, there were cattle gathered around puddles and drinking. Everyone began crying tears of joy and thankfulness. From that day onward, the region began to experience flooding. For our final event in Kenya, Tanya hiked five miles up a mountain to share Jesus with a remote village while being five months pregnant. We are not worried about another miscarriage because the due date is in exact time with the prophecy we received from Pastor Bill at Bethel. We will be flying back to Mozambique tomorrow.

TRANSFORMATION

December 14, 2009

Christian and his family left everything behind in America and live with us in Mozambique now. He is my best friend and we always laugh and have fun together. We all live together in the same little house. Christian and I working together always results in acceleration and expansion. The Kingdom of God is expanding in Niassa. Daily, souls are being saved from darkness, families are being transformed, miracles are occurring, and I am growing more tired. Christian and I like to launch off jumps and do tricks in the villages with our motor bikes. Everyone gathers to watch us, and then we preach the gospel. The children take turns preaching too. Maria asked me to teach her to drive my Quad in Assumani. I sat behind her and her thumb locked the throttle down as she froze in panic. I was unable to pry her thumb off from behind as we accelerated down the dusty road. People were shouting and jumping out of the way as we sped toward someone's courtyard. We plowed through the bamboo gate and into the side of a mud brick house, knocking a large hole into the wall. As we were launched off the quad, we flew to the side instead of forward so that we did not hit the house. I turned in the air so that we landed on my back. I felt the presence of angels lift and carry us in the air and we landed softly on the hard dirt. The bamboo fence cut Maria's mouth but she is healing. She is Phinehas's wife. Some men returned with me and repaired the wall and fence.

Our Bible school meets three days a week at the house we renovated. The men and women are growing more passionate and powerful in their faith every week. Some of them have already started churches in their own villages. When we come together, we all sing and dance, study the Word, and memorize scripture. I am so thankful. The victories do not come without much opposition. I often find curses from witchdoctors around our house. I noticed a rolled paper tied up inside our lemon tree. When I took it down, I discovered it was a curse written against my family. It was written in chicken blood with text taken from the Quran. I just laughed about it and set it on fire. Tanya is due to have our baby any day now. She has been doing well with her pregnancy here in Africa. We will not go to the hospital because it's not safe here. There is a Dutch midwife who's planning to help deliver the baby here at our house.

The Miracle of Zoe Life
May 22, 2010

Tanya gave birth on the night of December 23rd. The midwife finally arrived after I was in panic for a while because I didn't know what to do. We only had candlelight, buckets of water, and towels. It wasn't an easy birth. Zoe didn't want to come out. I had to keep pushing on Tanya's belly and after a scary panic, having to cut, reach in and pull, Zoe came out with the umbilical cord choking here around her neck. She was handed to me, blue and not breathing. Her head appeared deformed and I was terrified. We put her up to Tanya and she started to drink. Color came into her. Zoe is alive and well now. I am thankful that Debbie stayed on her knees, praying in our hallway the entire time of the delivery. Zoe is a miracle.

The following is Tanya's personal recount of Zoe's birth:

I went into labor the night of the 22nd. I had contractions all night long and the entire next day. It was 9:30 when I finally gave birth. Zoe hadn't been breathing well for about an hour or more after birth. It was about two hours after birth that they put her to my chest, after I had gotten sown up. she was on your chest first. Then I asked the midwife if she should be fed, since it was about two hours after birth. She said that Zoe would not have an appetite to eat because she had such a traumatic birth. Just then, Zoe started to suck on her hand. The midwife said that maybe she will latch on and eat, so we tried it and Zoe started drinking immediately. Color washed into her face right then and at that moment, I knew that she was totally healed from whatever happened to her in the birthing process. It took the midwife two days to be able to process and come back to try and explain what had happened. She said that the umbilical cord had broken and there was no reason for it. My baby was in perfect position to birth Zoe and there was no reason why she was not coming out. She said that she was not worried the baby would die since God was in control, but she was concerned that Zoe would be autistic since she had been without oxygen for so long.

Zoe's first words have been in the language of Yao and Portuguese. I love it! I often put her down for her nap by letting her ride in the carrier on my back while I drive my quad around. She goes to sleep so fast that way. People from far away villages are coming to see her every day. She is the cutest baby. I am so thankful for our cat today because it hissed and clawed toward Zoe when I held her down to say hi. It was right outside our front door and as soon as the cat hissed, I quickly pulled Zoe back and a black mamba snake struck right where Zoe's face was a moment before. This is why I am thankful for our cat today. The Yao people of Niassa are no longer an unreached people group. They were unreached when we first arrived here, but now the men and women who have graduated our Bible school are continuing to go to all the villages and everyone has heard about Jesus. Christian and I turned one of the walls in our house into a wall-o-map. It is eight feet by twelve feet and maps out the entire province. Most of the villages in this province are not located on any maps, so we bought these maps from the agricultural department, scouted and mapped out the entire province, and have marked them here on our wall-o-map. We are now preparing to return to Pemba to help lead the next Missionary Training School. We have helped lead a few of these schools so far. I get to meet a lot of amazing people from around the world when we're there, but it's not easy to take care of hundreds of people at one time. It helps me respect Heidi so much more as she takes care of thousands of people every day.

HOME INVASION

August 2, 2010

I turned 25 today! The Harvest Missions school is going extremely well. I've been meeting with a friend and planning a trip to Asia. He used to live in Nepal and has a lot of experience in that region of the world. We're going to start in Thailand and help lead a pioneering trip to start a new ministry base there. Graduation is next week and then we will be driving back to our home in Niassa. Our plan is to head to Asia in a couple of months from now.

When I teach here, the highlight for our students is when I tell the story of our recent home invasion of fire-army ants.

It was 10pm and I sat up in my bed because I felt waves of evil pouring into my house. In the spirit, I saw a large ant eater (Aardvark) in a robe, walking uprightly toward our home. I got out of my bed and prayed over our entire house until I sensed the peace of God return and the evil leave. At that same time the following night, we started hearing what sounded like rain inside of our house. "Maybe it's rats!" Tanya exclaimed. I turned my flashlight on and got out of bed. There were thousands of ants pouring into our house from under the doors, from beneath the windows, and falling from the ceiling rafters. They were already covering our mosquito net. I woke Christian up to help and we had the women and children move to the corner of the house with the least ants. We prayed, cursed, prayed, and tried not to curse while we stomped on ants and sprayed cans of raid. The ants were climbing up my legs and falling on my neck, and they were biting, and they were fire ants. We were losing the battle, so we moved our wives and children into the bed of my truck outside. When we shined our lights on the ground, it was swarming with millions of army ants coming from every direction toward our house. I declared, "I break this curse of witchcraft and I command all of you ants to leave my house right now, in the name of Jesus!" At that moment, all the ants reversed and the swarms begin to exit the house and start moving away in every direction. I went back inside and there were dead ants everywhere, so we went and slept on the floor in another missionary's house. We spent the next day cleaning up and picking a lot of ants out of our mosquito net.

We are yearning, longing, burning for our bridegroom King. He is ever-present yet He longs for so much more from us, to be completely united and given unto Him because His raging jealous love burns with desire for our full affection. He constantly calls us to come closer and dive deeper into devotion so that the fullness of His purpose for us as Sons, Daughters, and heirs can be fulfilled as we are sanctified and consecrated, set apart from the world to reveal and invite all into the Kingdom that cannot be shaken, to come to the desire of all nations; Christ Jesus our Savior King. We left home, family, control and comfort, to fulfill His purpose for our lives. This does not mean that somehow we have arrived and are super-Christians. We have so much further to press in to fully embrace Oneness with Him. He is perfectly holy and has no mixture in Him. So He calls us to cast off all compromise of conviction and despise the demon of mediocrity so that no veil may lay over our hearts and minds, but that with clear vision we may view from eternity's perspective, placing no value on temporary treasures that fade with the wind and are simply a deception of false security and comfort attempting to fill the void that aches for Him. Within our soul we grown for Him, so we naively numb the pain through seemingly innocent pastimes which seem good and are the enemy of the full eternal destiny that Love paid the price and sacrificed His life to offer freely to us. Few only want the crumbs and settle for less, disregarding the high cost of sacrifice that was paid on our behalf to redeem us from corruption and give us abundant life, which is so much more than forgiveness and a ticket to heaven. He is desperate for us to be desperate for Him and longing for us to long for Him. When we become truly hungry for the bread of heaven and crave and desire Him above any earthly thing, then we will begin to rise above every obstacle and fly on new wings with heavens perspective, and joy that vanquishes every opposition of dissolution that settles for less than ALWAYS MORE.

CHAPTER 6

Stepping onto the Waters

God, I pray for guidance and wisdom for this journey ahead. I am trusting you for the provision we need along the way and that you would help me lead those who are coming on the journey with us.

Thailand and India
February 11th, 2011

We've taken five other missionaries with us and traveled from Africa to Asia. We have a few countries on our radar for this trip, but only had enough money for a one-way ticket to Thailand. We joined a larger ministry team in Bangkok. There were sixteen of us working together to pioneer a new base. The vision and heart are for a base to be set up that missionaries can work from to help rescue women out of the sex trade industry. We ministered in the cities that are famous for being worldwide hubs of this problem.

Yonnie did an amazing job leading the trip. We co-led and Christian came to help as well. We love pioneering trips to scout out new places and lay foundations for other people to come. We were able to walk the streets where all the women are sold and we handed them roses with Gospel messages. We shared about Jesus and prayed for some of them, as well as for men who were there for sex tourism. We also got to eat a lot of fried bugs from the street vendors. While in Thailand, we got our visas and flights purchased for India. We made plans to stay with Shanti and Jonathan in Kolkata. They are friends of ours whom we met in while in Africa.

The city of Kolkata is very noisy. I am constantly hearing thousands of vehicles simultaneously blaring their horns. The streets are always crowded with cars, bikes, and cows, and it seems there are no rules. I love the food! Authentic Indian food is my new favorite. Our hosts have been very kind to us. They had us celebrate Holi with them yesterday. It is the Hindu festival of colors. We threw colored dye on each other until we all looked like rainbows. After that, we went on our rooftop where we through water balloons and dumped buckets of colored water on those passing by down below. Jonathan is a great teacher who loves to have fun. He is really into martial arts and also plays in a major Indian rock bank! We played fun practical jokes on random people. We threw mushy bananas into the ceiling fan at the hotel and watched it spread across the walls, then had to rush to the train station because the street riots were getting dangerously out of hand as the mobs moved closer, burning everything, and hurting people. We had just enough money to purchase flights to Kathmandu, Nepal and we fly out tomorrow.

India

Nepal and the Himalayas
March 13th, 2011

I met Joel at the airport when we arrived. He's 6'8", like me, and he used to play in the NBA, unlike me. He is awesome and loves to worship, pray for everybody, and go trekking to the remote places. We've been visiting the sacred Hindu temples. They're full of idols and paintings of Hindu gods. They are scary looking, but we just pray to Jesus and worship while in the temples. The culture is similar to that of India. There are a lot of tourists who come here to go trekking in the Himalayas.

Pashupati is my favorite place to worship in Kathmandu. It's considered the most holy, Hindu place in the country. It's on the river and lined with large altars where dead bodies are being burned, then swept into the water. People travel from far with their dying family members so that they may pass-on there, and have their most proper religious rituals performed on location. There are many rooms with families who are waiting for someone to die so that they will be there when it happens. I was able to go into a couple of the rooms and pray for those who were sick. I thought I was going to pray for the dead to rise since there are a lot of people dead and dying. I did try, but it seemed better to just be with their family members who are mourning, and to help bring comfort. Many children swim and play in the water where the ashes of bodies are being swept. Others are drinking the water. We were kicked out by the local Hindu police because we were talking about Jesus. There are several Sadhu priests who stay at the fertility temples. Each temple is only 6 ft. by 6ft. inside with a big male genitalia idol. We crammed inside one of them and asked if we could pray. They gave us permission and we started worshipping, praying in tongues, and laying hands on the Sadhu to pray for them. The Sadhu loved it, but when the local authorities heard, it led to us being escorted out.

We just returned from the mountains. My legs are so sore. We left Kathmandu six days ago on a crammed little bus that blared Hindi music in our ears all day. The bus ride only took about 20 hours. There was no space, so we all folded up like pretzels. We traveled down a narrow road on the side of a mountain with a sheer drop of hundreds of feet to the side. Sometimes it seemed like our tires hung halfway over the cliff as we slid in mud and bounced around. I could see Land Rovers and busses at the bottom of the cliff which had previously fallen off the road. We finally arrived in a little village and transferred into Land Rovers. For the next eight hours we drove over rocky terrain and through rivers. There was no road, but it appeared the drivers were accustomed to that type of travel. The Land Rovers took us as far as they could and then let us out to walk the rest of the way. Each of us carried a heavy hiking pack. Tanya carried Zoe in a baby carrier. My hiking pack did not have my personal belongings. It was an entire sound system built to show the Jesus film for up to 750 people at a time. It is solar powered and runs off a battery. Six hours later, we were still climbing a steep incline. "It's only a little further," we were constantly told. Tanya hyperventilated and had to breathe in a paper bag. Most of us thought we were going to die. Finally, we reached the top of the plateau after dark. We could see the snow-covered beautiful Himalayan mountain range in the beauty of the final setting of the sun. We ventured down into the valley and were able to find the floor or benches we were to sleep on. This only took a couple of hours more. I woke up to multiple pieces of raw meat strung above where I was laying. I was able to meet the Tibetan families during the next day and set up where we were to show the Jesus film in Nepali that night. They never had a movie up there before. Everybody gets scared and runs away when the snake slithers toward Jesus in the film. They think it's real! Many were saved and healed from sicknesses and disease that night. Very early the next morning, a group of us awoke and walked eight more hours to the next village in the mountains. We showed the Jesus film and many were saved and healed again. We slept in the dirt and walked back the next morning to where we had left the other half of our team. The journey down the mountain was a little easier. A girl on our team twisted her knee at the start of the walk, so Joel and I took turns giving her a piggyback ride all the way down the mountain. Some strong Nepali men and a donkey helped carry the extra baggage. Soon after we arrived back in Kathmandu, a powerful 9 magnitude earthquake shook Japan. I contacted Yonnie, who had led the Thailand trip, and we made plans to meet in Japan as soon as possible as an Iris Disaster Relief team. Tanya stayed with the team in Nepal and I flew to Tokyo to meet Yonnie and the others to co-lead the Iris Relief Japan trip.

bodies burning on the altars at Pashupati

Temples of the Sadhu Priests at Pashupati

Japan, Fukushima Disaster Relief
April 3rd, 2011

Every city along the coastline here is devastated. I saw a large boat on top of a tall building that had not collapsed from the tsunami. Everything around it was rubble. There are railroad tracks spiraled in a twisted form through the demolition of what was once a beautiful city. Little blue flags are staked everywhere to mark where dead bodies are. We have driven straight through the forbidden zone near Fukushima where the nuclear reactors were damaged. I love sitting with people in the shelters where hundreds of displaced families sleep on the floor. I also love going outside and playing soccer with the children. We make Starbucks coffee and prepare hot food packets at each shelter we visit. My heart flooded with sorrow as I spoke with one elderly man here. He is so sweet, but I can see the deep sadness in his eyes. His once beautiful home is completely decimated. Everything he had is gone. This is the same story for the thousands of men, women, and children who are bundled together in these evacuation centers. They are all grateful they still have their lives. Most everyone has loved ones and family members who were swept away by the tsunami. We have been praying with the people here. They are hungry for God and some have given their hearts to Jesus. I cry as I feel the hurt and loss of hundreds around me.

Matt and I are sharing a room while we are here in Japan. Last night, I put the Furious Love movie on my computer for us to watch together. The part of the movie came when Shanti begins to share about her work in India. I started to tell Matt about the time I just spent with her in India and at that very moment Shanti calls me through skype on my computer. When I answered it, she told me it was an accident. As soon as I heard her voice, the entire building began to rock back and forth and my computer flew onto the floor. Through a 7.6m aftershock, we scrambled to the door and made it down to the street. Everyone around us were leaving their homes and coming to the street. It was obvious there was still trauma from the recent 9m earthquake and tsunami. It seemed like we were all waiting for another tsunami wave to hit. We are all okay and there was no second tsunami. I'm heading back to Nepal in the morning.

The Great Wall and Across Tibet
April 27th, 2011

(above) Breakdancing on top of the Great Wall of China

Our adventure through China was very interesting. We had just enough money for our visas and flights. It was difficult when I first got to Beijing. No one knows any English and I don't speak Mandarin. No one from the airport even understood the words taxi or hotel. We are getting through it. It's a difficult language for me. Hiking on the Great Wall is beautiful. I did some breakdance poses to make fun photos and then climbed on the side of the wall. We visited Tiananmen Square and the Forbidden City. The coolest thing I ate in China is scorpion. They are still alive when you choose one in the marketplace. When I selected the one that I wanted, the vendor stuck it in hot oil, then handed it to me. I couldn't bring myself to try one of the sea horses. They are too cute and still alive until you pay for them. We caught a flight from Beijing to Xining which is on the edge of the Tibetan province. Some missionaries I met in Mozambique live there and we travelled across the Tibetan province together in 4x4's searching for a certain village they would like to move to.

On Easter morning we hiked up a hill to take the Lord's supper together. It was very cold and windy but still beautiful out. There was nothing but open country as far as I could see in every direction. There were no roads there. We drove over open land, mountains, and rivers, following a compass. Occasionally we ran into Tibetan nomads who make their encampments with white tents and care for herds of white Yak. They are so kind and served us Tibetan bread with warm salt-butter yak milk. This is their tea. I've learned to love it. You must be careful if you don't like it because as soon as your glass is halfway empty, they will quickly pour more of the salty yak milk and drop another cube of yak butter in your cup without warning. It's important to finish everything I'm handed in order to not be offensive. We came across a remote monastery hidden in a mountain cove. We spotted it by the tall, pointed temple tops from over the mountains. Two monks greeted us as we drove in. They always stayed in the temple while their wives and children stayed in a small home outside, caring for their goats. The monks were young men, around 30 years of age. They had long, straight, black hair and sparkles in their eyes. They are obviously rarely visited and never have foreigners in that place. After they showed us around their temple and introduced us to their gods, we asked if it would be ok for us to pray for them. They were overjoyed and declared, "we love to pray!, but we have to go our prayer room." They led us into a small room and closed the door. A large golden Buddha sat cross legged, filling up one side of the room. Incense was burning and smaller idols sat around the walls. We were given prayer shawls for our heads and one of the monks took his sticks and positioned himself to play the large leather drum that hung down and filled the middle of the room. He began beating on the drum as the two of them chanted in Buddhist meditation tongues. I didn't know what to do. I didn't know if it was right for me to put the shawl on my head out of respect or not, and it felt very dark and sticky in there. I started praying and singing in the tongues of my spirit and laid hands on the monks to bless them. They immediately stopped their prayers and bowed their heads. We shared with them about Jesus the Messiah. One of the men believed in Jesus and wanted to know Him more. "I have never felt this presence before in my life and it's what I've always been searching for," He exclaimed! They begged us to return to them and teach them more. They needed Jesus more than anything else. I continued the journey with the rest of our team, stopping at the sacred high place which are decorated with Buddhist prayer flags. We worshipped Jesus at those high places, discovered the village on the western edge of Tibet that we had set out looking for, then returned to Xining.

We've returned to Bangkok and the others from our small team are beginning to fly home. I am not. I'm not sure what home is right now. I am taking my family and flying to Australia and Indonesia next. We've been invited by my friend Matt, who I was just in Japan with. The adventure continues…

THE OUTBACK

May 30th, 2011

Our time in Australia was awesome. We flew into Darwin and spent our time in the Outback with the Aboriginal people. They have lovely hearts and are very spiritual. I went fishing at the edge of a crocodile swamp where two rivers meet. The Aboriginal man's wife stood on the shore repeatedly yelling, "Crocodile!." Every time we pulled in a fish, we had to walk a few hundred feet back through the swamp to land to leave the fish and then return to get another one. I asked Matt what I should do if I see a crocodile and he told me that he would throw his pole at it and run, so that's what I was prepared to do. I didn't see any crocodiles there. We caught more than sixty barramundi fish in about two hours. They were all big. It was too much for us, so we fed the village when we returned. We also went crabbing in the mangrove at low tide. One of the local boys showed me how to knock their claws off with a stick before picking them up and putting them in the bucket. We had so many king crabs. We made a fire just offshore and ate crab until it was coming out our ears. There were plenty leftover to feed others.

The most exciting experience I had there was when we went fishing far out in the bush at one man's billabong. The landowner warned us to be careful and explained how his niece was eaten by a crocodile in that very location two years earlier. I had only caught one catfish and the sun was setting. Tanya, Zoe and I were at a separate location than the rest of the men and when I cast my line one final time, I got snagged on something near the other side. When I tried to snap the line, I noticed it would slowly drag toward me. I held my pole over my shoulder and stepped strong and steadily away from the water. When the line was just near the bank on my side, it snapped. I ran over and looked in the water. There were bubbles and whirlpools and I could see the large body of a fish moving under water. I took Tanya's pole and dropped the line in right above it. I was going to rehook this fish and have something to brag about! It wasn't working and it was still getting darker. I got down on my hands and knees, peering into the water. Suddenly the large head of a crocodile surfaced right in my face. It looked like a devil! I yelled for Tanya to grab Zoe, threw down my pole, and we ran back to the other men. We all came back together and could see the track of how the crocodile came out of the water right where we were and had gone back in another place. My catfish was still there, so that was good anyway.

We came from Darwin to Indonesia to minister with Matt. The local people farm and roast coffee here on the island of Lombok and it's so good! We are heading back to Bali today and then flying back to Africa. This is a very beautiful part of the world. I love traveling and I want to go to every country in the world. The more I travel, the smaller the world becomes. Being on the move constantly is becoming normal for me. I love experiencing new cultures, languages, and geographic regions. We will be helping lead the next missionary training school in Pemba, Mozambique. I hope that our Yao friends are doing well and continuing strong in the Lord. We will only have a brief time to be with them at our home in Niassa.

Faith provoking risky action is a key ingredient in fulfilling your humanly impossible dreams.

Impossibility is non-existent within the vision of the heart's eyes that rest in the realm of the King's domain.

The noise and glare of intimidation from Satan's raging hordes attempts to quench the flame in the dreamer's heart.

But the roar of hell is laughed away by the heir who's aware of their inheritance in Christ.

Once awareness of identity is fully realized, then the revelation that all creation has been groaning for begins to burst forth with the power of spring life, and the full manifestation of redemption shines from the dawn horizon.

CHAPTER 7

The Glory Trip

An Impossible Journey

September 14ᵗʰ, 2011

We spent the summer back in Africa helping lead the Harvest School of Missions and experienced a shift that opened a door of impossible opportunity for us. What lies before us now is an adventure and challenge greater than anything I've experienced thus far. Benjamin Gomez is one of my friends from discipleship in L.A., and he is now a missionary in Peru. It has been stirring in my heart to visit him and support his ministry. As I considered the option of driving from the U.S. down to Peru, I decided to post a notice on the gate to our housing area. The notice read, "If you have a heart for South America, please come to our house for coffee on Sunday morning at 10am". I had to make a lot of coffee because fifty students showed up that morning. We continued to meet every Sunday to pray and dream together. The vision expanded as we wrote down everything the Lord was speaking to us. We will drive from the PNW, down to the tip of South America along the west coast and drive all the way back up the east coast. We will carry the fire and glory of God to make a ring of fire around the entire continent. We will go to every country in Latin America, picking people up and leaving people behind along the way. We will carry the love of Jesus to the jungles, the remote villages, the churches, the slums, the rich and the poor. Many are excited to be part of the journey which we are calling a "Glory Trip."

I brought the vision to Heidi and she was excited and supportive. She promised to help in any way that she could. Iris Global is going to bless, cover, and help promote our mission. After graduation in August, we returned to America to prepare for the Latin America Glory Trip. People flew into Portland from around the world to help us get ready.

Everyone joining us to embark on this glory trip has given a deposit and is expected to contribute $300 monthly to the team fund. This way, we can have gas money and team meals along the way. We have pooled together all the initial costs paid and bought four vehicles and two pop-up campers. We have a heavy-duty Suburban that we named Overflow, pulling a trailer named Shalom. We named our Durango, Shekinah, and it pulls a camper named Glory. We have one motorhome van named New Wine and our motorhome that Tanya and I purchased last year as a promise from God. We named it Open Heaven and she is a 1982 class-C beast. Christian spent the last year refurbishing it for us while we've been away. Taylor will be driving from Texas and meeting us in Tucson, Arizona. He has a Buick station wagon which he named Counsel and Might. We've purchased and gathered camping, cooking, and living supplies and packed them into the vehicles. We're continually warned from others not to go near the Mexico border right now because a violent drug war has broken out and it's not safe. Many who live near the border towns are fleeing to other parts of the country and some borders have closed. We are going anyway.

We've now driven from Vancouver, WA to Tucson, AZ and are heading to the Mexico border tomorrow. Our first stop was at Bethel Church in Redding, CA. Heidi Baker happened to be there at the same time and our team was announced and called to the front. The church prayed for us, commissioned us, and sent us out with a generous offering on September 11. There are twenty-two of us on this team now ready to make our first border crossing together.

DANGEROUS ROADS

November 9, 2011

We've been together on the road for almost two months. Our time has not been without dangerous challenges mixed with miracles and glory. It's after 2am and I just got everyone situated into rooms in this broken type of motel. We are somewhere lost in Guatemala. We only arrived here by a miracle… many miracles, just today. I had a destination in mind planned out. Our GPS directed us to a certain small town and then it was somewhat confused about where to get onto the next road. We asked many of the locals for directions and we were instructed to take the road that exited through the cornfields. After we left the town, there were occasional bystanders on the road who all urged us to turn around. "This is not a safe road!", they warned us. We were told to turn back or we would surely be robbed and killed up the way. That made it scary, but we could not turn around anywhere. There was a sheer drop to one side of this narrow dirt road and the other side was a wall going upward. We would have had to back our five-vehicle caravan with trailers all the way back down the mountain. So we took our chances going forward.

It became very dark outside and I could no longer see three of the vehicles from our caravan. I stopped and waited for a while, and still nothing. I decided to run back down the mountain on foot since we could not turn around. I found the stranded vehicles around a few bends. The tire from the Shalom trailer had blown. The rim was beaten out of shape and the tire badly damaged. Earlier that day, this trailer had already blown a tire. This was now the spare, and the original tire had not been fixed. A couple of the guys relentlessly tried to fix it with a rock and a string. I headed back up the hill on foot to inform the others what was happening. About halfway, several large angry black beasts came out of the bush with eyes glowing and chased me, growling and nipping at my feet as I ran for my life. I managed to leap onto the side of Open Heaven and climb up for safety. I helped unhook Shekinah from Glory and the small Durango was able to go back downhill to the others. They seemed to be happy and having fun when we got there. We were going to have to roll the trailer off the cliff because we had run out of options. Taylor was still determined and not giving up. A few minutes later he declared, "there!", and stepped away from the trailer which had the tire on, fixed and ready to ride. They had done this with only a rock and a string. The photos the girls took of them working on it clearly show angels with them working on it too.

(above) Taylor receives angelic help in an impossible situation.

Again, I lost a couple of vehicles from the caravan and had not seen their headlights for a while. I waited, and finally one of the women came running up the road shouting, "Come quick, Overflow is falling off a cliff!" I went very quickly because Zoe was in Overflow. The Suburban and trailer were on a steep and slippery hill and the trailer was already halfway off a very large cliff! This was not an easy task, but somehow we managed to unhook the trailer, push Overflow back onto the road, and up the hill to where there was traction again. Then we managed to push and pull Shalom back off the cliff edge and up the steep hill to reconnect it to Overflow. At this moment, I was thinking that maybe we should have pushed it over the cliff earlier that day. This was also only handled with the help of angels and many strong men working together.

Everything was flying out of the cupboards of Open Heaven and hurting people as we bounced back and forth on the road. Often the road became narrower than my motorhome. This was so evident at one point that I had everyone get out and walk while I closed my eyes, prayed in tongues, and drove forward. And then we were on the other side. At one point we came to a muddy river running through the road. There was a steep ledge on each side. I looked at it for a while, then got back in the driver's seat. We had only one option, onward. I plunged through it first, down in, then bounced back up on the other side. Each vehicle followed and successfully made it through. We were all stopped on the other side, gloating on how we made it through when Simba paused and said, "What's that smell?" I inspected the vehicles and discovered that the sewage tank of my motorhome had completely exploded when I plowed through the muddy river before the others. Everyone uses the toilet in my motorhome while we are on the road. Now, the four vehicles that followed me are all covered in the contents of our septic. Cool! We continued through the night until we finally arrived here. I don't know where "here" is, but we are all thankful that we are alive and made it to somewhere. I am sure we will figure out where we are in the morning. The GPS hasn't worked at all out here.

Christmas in Colombia
December 18th, 2011

Today was our first day in South America and I got lost. Well, I still am lost but now I at least have made contact with the rest of the team through the internet. We arrived here in Cartagena, Colombia and got settled into the YWAM base where we'll be staying. A bunch of us went to the beach and I went to grab sodas. When I returned to find the others, they were not there. I walked up and down the beach so many times looking for them, and it seemed they just disappeared. Now that I'm talking to them, we realize there was just some confusion and they are back at the base. My new Colombian friend will come get me soon because this is not a safe place. Our vehicles are still waiting to board a barge-ship from Panama. They should be arriving in a few days. Most of our team wanted to drive through the Darian Gap. We did a lot of research and found that it was not a possibility. Still, a group of seven were determined to cross the Darian Gap, which is the most dangerous location in the Western Hemisphere. They made every effort possible, while some of the older members of our team were insistent that I stop them because they would die. I supported them one hundred percent. I, personally, only couldn't go because I need to take care of Zoe. Our team is our family. They are hardcore for Jesus to the point of being ready to give their lives for Him in a second. They knew that they would probably be captured by the guerilla soldiers in the Darian Gap. This made them excited because then they would become missionary prisoners. Every attempt in the Darian Gap effort failed. When we got to the shipping port in Panama, we were giving the lowest price possible of $12,500 to ship all the vehicles from there to Cartagena. None of us had more than a couple hundred bucks. When we put all our money together, it was not even half as much as we needed. We all prayed and stayed there waiting. One of my young team members from Australia took me aside and informed me that he had the full amount and would pay it, as long as I do not let anyone know it was from him. So, we did that and here we are. The vehicles will be arriving soon.

Ecuador- Danger in the Amazon
February 20, 2012

The sun is setting now and I'm on the top of a small mountain by myself. I hiked up here to see if I could find reception for my phone and I managed to get one bar. It was just enough to call my mom and let her know we may never be heard from again. I need to get back down to the rest of my group in a hurry because they're all terrified. Our team split in half back in Quito and I brought a group to the southern portion of the Amazon region of Ecuador. We wanted to find the Waorani Indian tribe. The Waorani tribe of the Northern region have been reached with the Gospel after Jim Elliot and his friends were speared to death when first making contact with them. We were told that they were still unreached and very hard to find here in the southern region. Our first few attempts failed as we were forbidden by a large oil company to enter the area. We had given up and then we met a young man in the marketplace who is Waorani. He went by the name of Marcus. He spoke fluent Spanish because he had come to the city to go to school. His uncle was the king in a certain territory deep in the jungle and Marcus would take us to him. After a long journey to nowhere, we arrived at his uncle's village. They were shocked to see foreigners come way out here, and they received us warmly. The Waorani tribe painted our faces for war, then taught us how to throw spears and shoot blow darts at birds. Tanaka, the king, told us how the last time missionaries came to their area, the husband and wife were killed by the Tigris headhunters, and their daughter was taken captive. Marcus was translating for his uncle, but nervously hesitant to let us know what his uncle spoke. The Tigris are Waorani Indians who separated themselves deeper into the jungle because they hate civilization. They are enemies now because Tanaka and his family sometimes wear t-shirts or have used bar soap. Tanaka shared a little about growing up with his father in the jungle. Neighboring tribes would often come and spear everyone where they lived and they barely escaped. He and his father and uncles would seek them out and spear entire villages of neighboring communities. He had grown up in a jungle life of war and spears. We had our tents set up in their large thatched hut. We arrived late yesterday evening and today seemed to go okay, until now. As Tanaka continued to share, most of the young women who were with me became wide-eyed and shaky. Tanaka told us how the Tigris had let us know they are nearby by leaving their footprints last night outside our hut. This is what they always due to signify they will be returning the following night to kill everyone. It was already evening and we had no way to leave. This was normal life for Tanaka and his family, but I think that we've gotten ourselves into more than we were expecting to. One of our young ladies asked Tanaka what would happen if they came tonight. Tanaka somberly replied, "I will defend you, but they will kill my family and me, and then they will kill all of you, and they will take Zoe alive as their captive prize". Tanaka was very serious about all of this. This is the only life he knew and he expected the Tigris to arrive that night. This is why I've come up to this mountain top to let someone else know what became of us. I am thankful to God that I found one bar of cell phone reception.

I don't expect anyone will sleep well here tonight. This jungle is full of strange sounds and my team thinks we're all going to die. I plan to stay up all night praying and standing watch outside. One cool thing is that we were each given a Waorani jungle name and got to eat strange foods here.

Peru- Lost in the Amazon
March 27th, 2012

We are now deep in the Amazonian region of Peru. We found a bus to take us from the coastal region straight up and over the Andes mountain range and down here to the Amazon. There is a famous terrorist organization who hold power in the mountains and we were warned not to go, but we're okay. There was an armed officer who boarded our bus somewhere in the mountains to let us know he was there to protect us, and he asked for money. Everyone got extremely sick from the rapid rise in elevation. We went from 0 ft to over 22,000 feet in a few hours. Our bus was not allowed to stop, and windows could not be rolled down. We all survived with a lot of moaning, laughing, and vomiting. We spent the next couple days with hammocks on boats traveling down the Amazon tributaries. We're here with the aim to find the Ashakani people. They are an entirely unreached people group hidden deep in the jungle. The sunrises and sunsets here are the most beautiful I've ever seen and pink dolphins often follow our boat, jumping alongside us. We finally arrived in this remote village and camped for a night. Yesterday we boarded two canoes with about 15 of us in each one. Our pastor friend operated our canoe while a guide was navigating the other. There are no maps for out here. They only know their way based on fishing trips where the guide once made contact near the Ashakani people. The entire jungle is flooded, which means there is no land to stop at after this village. It is just thick jungle coming up out of the water. After hours in the canoes, the channels became smaller and smaller with countless divides where the pastor and the guide would argue about which route to take. Finally, our canoes divided into separate channels. Hours later, we were lost, hot, hungry, and having to pee off the side of the canoe. We kept praying and worshipping until we noticed that at every divide, there were a chain of golden butterflies hovering over only one of the two channels. We started to follow the golden butterflies until evening. Then we heard a faint voice in the distance. I called out our team codeword, "Shabba"! "Shika Baba!", someone shouted from the distance. We continued these calls until eventually we found the other half of our team. They were excited because they had come very close to the Ashakani people and met a couple men from that village who were fishing. The men forbid them to come to their village because the only other time foreigners had been there, they had stolen their children and harvested their organs to sell. Our team insisted with them that they wanted to go anyway, even if they are killed because they want to share a message of love with their people. Our guide had refused because his wife and child were in the canoe with them and they would all be killed. A giant rainbow filled the sky and God promised that the Ashakani tribe would be reached with the Gospel.

We were thankful to be reunited, but no one knew how to get back to the village where we'd stayed the night before. The pastor had been asking me for directions most of the day now because he had no idea. It was dark by this time and as the channels got wider we decided that whatever direction the water was flowing faster, we would take going back because it would eventually lead to the main tributary. We often ran into the side of the jungle wall from not being able to turn fast enough. Large spiders would fall on us, but this was not as scary as the anaconda we had seen and the many crocodiles floating by as they stared at us with their glowing eyes. We often nearly tipped over, but we had many more angels with us because Zoe was in our boat. I stood on the front tip of the canoe with a dying flashlight. I shouted back to the driver which way to turn at each divide. We all lived and are back in this village now. The people gathered in their schoolhouse earlier this evening to hear from us why we came. There were two witches present and at the end of our meeting, everyone received Jesus, many were healed, and the entire village prayed a prayer with me to repent of witchcraft. Now we are just trying to sleep on this wood floor, despite the swarms of gnats, and the rats trying to nibble our toes. I'm kind of excited to get back in our hammocks on a boat again. We are all covered in red bumps and itching from the gnats, but the piranha tasted so good!

(above) Zoe and her daddy in Bolivia

My Personal Struggle
April 2nd, 2012

Machu Picchu is one of the New Seven Wonders of the World and was well worth the detour for us to visit today. Christian was able to fly down for a couple weeks and rejoin us for this part of our Peru adventure. He helped us set up this entire trip and co-led with us until El Salvador. At that point, he had to return to his family in Washington. My team is concerned about me, which I think is part of the reason we all helped bring Christian back for a bit. I'm tired and I've been drinking alcohol more as the journey continues. I initially set a rule for our team that alcohol would only be allowed when we are not being hosted by churches and not in ministry settings. After Christian first left, I began feeling lonelier and isolating when I had opportunities. Tanya is always here, but she is very busy pastoring our group. Everyone on our team is amazing, but somehow I am still feeling alone. I often drive multiple 14-hour days in a row, and I never let anyone else drive my motorhome. At the end of every day, I make sure everyone has food and a place to sleep. This is often not done until 2am. When everyone is taken care of, I isolate, drink, and sulk until I can pass out. People are continuing to be saved, healed, and delivered everywhere we go. Due to internalizing all the intense stress and pressure from leading, I fried some circuits in my brain. I am still a strong leader and determined to complete the task at hand. I am not sure if I'm okay anymore.

The End of the World

May 22, 2012

The City of Ushuaia is on the island of Tierra del Fuego at the tip of Argentina. Ushuaia is also known as, "The End of the World" because there is no further city south in the entire world. We've arrived at the end of the world now. When we reached Punta Arenas, we had all run out of money. It is the city at the southern tip of Chile, just north of here. We arrived there in the cold of night and prayed together for direction. We could not sleep in our campers because we left our vehicles in Santiago, Chile. A couple of our team members went walking into town. They came back within twenty-five minutes and had arranged our stay at a hostel for only $2 a person, per night. The next morning we heard worship music playing loudly from across the street and Taylor went over to explore where it was coming from. He came back and testified how a church had invited him to the front and blessed him and they were now requesting our entire team to come that evening. The church gracefully took us in and hosted us in their households. They gave us all food, shoes, and money, and then sent us on our way. Our bus did not pull into Ushuaia until 9pm and we didn't know where we were going to stay. We formed a large circle outside the bus as we stepped onto the End of the World. As we began praising the Lord, an older gentleman hobbled over to us and introduced himself. He let us know that he had been waiting for us, had places for us to stay, and he had a lot of people and dinner waiting for us at his church. We didn't know who he was. He had received word from someone that we would be coming. We followed him to the church and ate a wonderful meal. Now we are unpacked and settle into a comfortable place that was all prepared for us.

Despite all the miracles and goodness of God being poured over us everywhere we go, I'm continuing to drink way more than I should whenever I can get alone and take a break. I feel like I'm pretty good at hiding it from most, but I'm probably not. People are worried about me.

At World's End

Buenos Aires

June 14ᵗʰ, 2012

Buenos Aires is a big city! Our pastor contact was supposed to meet us when we got into town, but it appears that we were stood up. We drove around for hours trying to find a place to stay until we finally stopped at a grass lawn in the middle of two gang neighborhoods and set up our campers and tents. Early in the morning, the police were banging on our camper door to let us know that we needed to bring our car battery inside or it would be stolen. They had already made those sleeping outside pack up their tents and belongings. "It's not safe here", they warned us and told us to leave. So yesterday we drove into town until we found Claudio Freidzon's church. Heidi had told me to visit him if we came to Buenos Aires. He is known to have been helping lead an incredible movement of God in this part of the world. During the service, he publicly recognized our team and prayed over us. After the service ended, we were informed that he had paid for our entire team to be put up in a very nice 4-star hotel for the next four nights. It is luxurious here, especially in comparison to where we slept the night before. Tanya is pregnant and it's nice for her to have a break from the holy chaos of this incredible journey. I am still drinking alcohol after everyone turns in for the night. It is the only thing I know to be able to cope with everything right now. I'm carrying more than I'm able and I drink to forget about it until I pass out.

Fortaleza, Brazil

November 10th, 2012

I've spoken with the rest of the team about what we should do with our vehicles. We can't drive them any further north because we are just south of the Brazilian Amazon. Everyone prefers to bring all these vehicles back to America. We've become very attached to them through the journey. We've been traveling together for a year and a half now in Latin America. It's too expensive to ship them all back to America, so we are donating them to our ministry friends here in Fortaleza. Tanya is nine months pregnant, but after we looked into getting a passport for the newborn, we found that it would be impossible in this city and very complicated in this country. I asked God for four confirmations if we were supposed to fly back to America for the birth. All four confirmations came to pass in one day. Most of our team have already left ahead of us to the Amazon jungle, and only a few stayed behind to help Tanya in her pregnancy. I don't know what I am doing but I still go out and drink a lot by myself, then come back and wait. We are getting on a plane back to Washington in the morning. You're not permitted to fly if you are over 8 months pregnant. Tanya is going to hold a pillow over her stomach when we check in and hope they don't notice. My biggest fear would be that she goes into labor on a plane, but with these four confirmations from God, I'm confident enough to go to the airport. I will miss the last three countries of Northern, South America and all the Caribbean, but a portion of our team will continue forward all the way to Florida. I can come back and finish later. My son Zion, in Tanya's belly, is more important right now.

I am grateful for the men and women who have continued strong with me throughout this Glory Trip. We've become a close-knit family and share all things in common. I became a wreck and often appeared drunk around my team. We occasional appointed new co-leaders, and more recently, new leaders. My core group called me to meet with them a few weeks ago and told me honestly how my failure was affecting all of them. As a leader, all my victories and my failures affect everyone who is following me. I have failed and failed a lot. I am burnt out, tired, and too prideful to stop. No matter what the cost, I had to complete the mission until the entire vision was fulfilled. It has been fulfilled. It was not by my good leadership or my own ability, but by the faithfulness of God and the unified support of my team, whom have become a family in the purest form.

Faith is unstoppable

Does not acknowledge the impossible

Considers the onslaught of our enemy as merely laughable

Sees difficulty as opportunity

Runs into challenges, springing off of them like a trampoline

walks through walls of the limitations in the minds of men

A man of faith may fall seven times, but rises again

Because faith has substance that is currency in the spirit realm

It's through faith that we are born-again, receive the Spirit and live for Him.

Faith sees the unseen, receives promises as guarantees

Looks ridiculous to the world because to our human reasoning it never agrees.

CHAPTER 8

Micronesia – Level 2

ZION SHONE FORTH!

March 11th, 2013

Thankfully, Tanya didn't go into labor during our flight and Zion was born in my mom's bathtub a few days after we returned. Zion grew in Tanya's belly during nine months of extensive travels through South America and was born a handsome, healthy boy on November 14th. One month after his birth, the biggest storm of 2012 hit and decimated the island of Mindanao in the Philippines. I felt led to go and bring relief to the victims. Christian and I organized a disaster relief trip and flew out to spend Christmas on the island of Mindanao. Most of the Filipino people have beautiful hearts to glorify and thank God in the midst of devastating loss. Large crowds of people turned to Jesus, were healed from sicknesses, and filled with the Holy Spirit in every place we went. Christian and I assembled over 200 Christmas gift bags to pass out at the evacuee centers. At the last shelter, I counted the remaining gift bags and there were 34 left. The administration showed us that they had forty families staying there on their list. I recounted with the pastor and confirmed that we only had 34 gift bags. Faith filled my heart and I looked at the pastor and said, "It will be more than enough". When we gathered inside, the staff marked each family off their list as they came forward to receive a gift. All forty families received one gift bag each and we still had ten left over. God had multiplied the Christmas gift bags! I returned to my family in Washington and began preparing for our next big adventure.

Christian and his family had served in the Salvation Army fifteen years ago in Pohnpei, Micronesia. He was led to return to Micronesia and Heidi gave us approval to start a new Iris base. I decided to move my family and join him to help establish Iris Micronesia. I heard about how beautifully majestic it is and I love tropical islands, so I became very excited.

We've been on the island of Pohnpei for a few weeks now. It's the most beautiful place in the world that I've ever seen in my life. It is mysterious and resembles the fluorescent landscapes from the movie Avatar. We have a house next door to Christian and the Ocean is across the street. We've been given approval to start the first Level-2 Harvest School of Missions here. This means the school will be primarily for those who graduated in Mozambique or who have already been serving on the mission field. People from around the world are already submitting applications. Tanya and I will be flying to Mozambique soon to direct the next missionary training school while Christian stays here and does the groundwork for the Level-2 school. We'll be returning the day after graduation in Mozambique and the students will start arriving for the Level-2 school the following day. We never have a moment to rest anymore.

(left) Christian Jung (right) Jesse Gellatly

Awaken the armies

It's time for the final battle

Shake off the yesterdays because today is like no other

What's coming now cannot be expected because something's coming, something new under the sun

A tune is humming that has not been heard before

The final showdown, a power conflict like no other

The sleeping masses will be awoken with a startle

Can you feel it? The atmosphere is vibrating with anticipation and the earth is groaning as it feels the day coming.

Prepare for war. Your victory is secured, and the time for redemption of the purchased possession draws near.

Missions Movement

July 21, 2013

We've been back in Africa for a couple months. After we landed in South Africa, Tanya and the kids flew to Pemba and I flew to Niassa. I intended to pick up our truck and quad and drive them across the top of the country so that we would have them during our time in Pemba. When I arrived in Niassa, Faithful's axel had been broken and was still not fixed. Through continuous failed attempts and time spent with the mechanic, it was still not fixed and time had run out. The school that I was supposed to be directing in Pemba was starting without me. I caught a minibus to begin the three-day journey over. I was very sick when I arrived in Pemba and went straight to bed. People assisted me to get out of bed and go to the nurse's station, only to find out I had typhoid fever. I spent the next ten days in bed. We had twenty-two strong staff members helping lead the school and sixteen of them were members from our South America team. We were not alone and the school was going incredible even while I was stuck in

71

bed.

There is a pressure to shut down our base and upcoming school in Micronesia, so I'm torn between two places and frequently on skype calls. Christian made a mistake six month ago that just resurfaced and became public. Now he's being forbidden to have anything to do with our new Iris base and the Level-2 school. Christian was the primary pioneer for all the work in Micronesia, which made everything very difficult at this point with him being taken out of the picture. I convinced Heidi that we must continue with the Level-2 school regardless of everything going on and I stood in the gap for Christian. I continued to defend him when countless voices slandered him with ridicule and shame. A group of our staff are arriving in Micronesia early so that Christian can be begin handing everything off to them. I can't stop now. Many students have already paid their tuition and have their flights booked to Pohnpei. We have well-known guest speakers scheduled to come and teach and we've rented the nicest beach/Marine Park on the island for three months to host the school. We will have 64 international students and 36 Micronesian students for this first Level-2 training school. This is a major obstacle to overcome but I am determined to keep going no matter what stands in my way. We will be teaching the students Bear Grylls type survival training, spearfishing, cross-cultural living, and Holy Ghost empowerment. I must continue to see this to fruition no matter the cost.

Nearly Drowned
October 10, 2013

It's 2am and I just returned to shore, thankful to be alive. I got dried off while the fish were being unloaded. We were spearfishing a mile out at the barrier reef with a group of students and locals, as we do at least once a week here. It's stormy tonight. The waves were ruff and the rain poured. When you're face down in the water, you don't notice it much, except for the tossing waves on the surface. It can be freaky out there because there are often reef sharks, sting rays, and dangerous eels. I can only see what's in front of my flashlight when I'm under water, so whatever is in the dark that I can't see is what I am afraid of. I like to stay in a pair with someone who is more experienced so that I feel safer. The boat drops us in pairs at different locations and then continues back and forth to check on us over a couple mile distance. I was following Anthony, who is known as one of the best spearfishing men in Micronesia. I have learned to free dive up to 30 feet, but he regularly goes to a depth of 100 ft to get the bigger fish. I came up to clear my snorkel and when I looked around, I could not find Anthony. I could not find our boat, and when I went back under water, I could not find the reef. The waves were high and it was still storming. It was just me and the ocean deep, in the middle of the night, and I was out of breath. I kept swimming in different directions trying to find Anthony or the reef but was unsuccessful. I was spent and could hardly stay afloat anymore as my fins kept falling off. "This is it", I said to myself, "I am about to drown and my body will probably never be found". At that moment I dimly saw the flicker of a flashlight in the distance. I dove down and moved as fast as I could in that direction. I would occasionally come up and take my mask off to look for the light again. It would take some time because Anthony spends way more time in the deep then he does at the surface of the water. I finally found Anthony and he was near where the reef is close enough to the surface to stand out of water. I took a break and eventually our boat came back by, so I got in and called it a night. My favorite part is when we each choose one fresh-caught fish out of our cooler, scale and gut it with our fingers, then eat it while we ride back to the island. Christian usually comes with us during these times. Certain leaders we assigned to the school hate it if he is around at all, but he's my best friend and loves to spearfish. It's hard for him to stay distant when we are all staying on a very small island in the middle of the Pacific together.

The main goal of this Level-2 school is missions mobilization. Many students who graduated Harvest School in Mozambique had a calling and desire to be serving overseas but didn't know how to get from A to B. We worked with each of them individually to get connected with people in the place their hearts burned for. We have placed them on different teams called to those specific areas and types of ministries where they will be sent directly after graduation on their long-term outreach. Many of them will stay where they are going and plan on living there to begin their life as a missionary. Personally, I had been doing very good for a while. I'm always extremely busy and I coordinate and manage everything nicely. I've started drinking again but I try to hide it well. I often stop off at a beautiful, isolated cliff-edge location by myself and down a few bottles when I have a break and then I go back to taking care of everybody else. Tanya and I will be leading a team of seventeen of our students on a Europe Glory Trip. We fly out the day after graduation and we already have Eurail train-passes, which gives us a limitless first-class train pass for the entire European Union. We have friends and contacts throughout Europe and most of the students going with us have a heart for a specific country there. I want to go to every country in the entire world and this seems like the best way to do it. We've become accustomed to always moving and having Zoe and now Zion with us as we go.

I was confronted by a group of our staff about being intoxicated last night. I denied it. I think that I cover it well and sometimes even deny it to myself. I must be stupid because I live on a tiny island amid a crowd of Spirit-filled people, and I'm the leader. I am 6'8" and think I can somehow hide so that people won't notice I'm really struggling. I miss Christian, but he is also struggling. I need to keep going. I don't have time to take a break or rest. I don't ever have time to take care of myself. I need to continue carrying the world on my shoulders because I think that I can do everything better than anybody else. Maybe I'm wrong. I wrote the following words and declared them over our students as I encouraged them to trample their fears.

Born to be wild

Trampling over fear with the untamed faith of a child

These Jesus freaks freak out the freaking devil with their fearless faith and Holy Ghost fire

You are the wild ones with the big guns

Ambushing the enemy's camps to recover the King's stolen jewels

To the world appearing as fools

But you are eternities heroes

You are unstoppable, blazing with fire unquenchable

Laughing in the faith of the darkest evil

Flying on the wings of freedom

Demolishing the forms built by the religious traditions of men

You are unexplainable, unconventional, practically irrational

And the whole earth groans for you

to be who you are...

A fearless lover, pulling the lost children home to the Father

CHAPTER 9

Europe Glory Infusion

Mandatory Break
November 4, 2013

I am currently in the Netherlands with my family, staying across the river from Belgium. I was placed on a mandatory rest/break and instructed to stay at this House of Prayer until I'm released to rejoin my team. We were put up in a nice cabin near Amsterdam when we first arrived. I was invited and hosted by my friend Mattheus van der Steen for a large global event called Mission Possible. The entire conference was televised worldwide with many national Christian leaders attending. There were leaders from the fifty most closed nations in the world there to represent their countries. I had the privilege of representing Micronesia and speaking to everyone at the conference about my heart for the unreached. My leaders from Iris met with me privately to discuss concerns brought to them about my alcohol use. I defended myself well and agreed to take a short break before meeting back up with my team in Europe. While I was lying on the floor during worship at the conference, I fell asleep. When I woke, I had no feeling left in my right foot. I continue to have drop foot today which makes the idea of backpacking Europe a little challenging. I love challenges.

FALLING APART
December 17, 2013

I am back in the Netherlands at the House of Prayer again. I had only stayed here about a week the first time and then Tanya and I were able to lead our team through 15 countries across Europe. We'd asked Melissa to co-lead with us, as she had also become our co-leaders for the latter portion of our Latin America Glory Trip. I could tell she was becoming worried and concerned for me throughout the journey because it was evident to her I had started drinking again. I somehow managed to figure out how to take care of everyone, preach in churches, heal the sick, and still find time to hide and drink by myself every chance I got. We were asked to return here and let the team continue with Melissa leading them. It's because of me. I am obviously an embarrassment to those who follow me when I drink. I have MRSA and open wounds have broken out all over my body. My fingernails are starting to fall off and I still have drop foot. I am sleeping in a bed alone because I stain everything with blood. It's very hard to find an escape here to be able to go and drink, but I try.

Mandatory Healing Time
March 2, 2014

We had been serving on the overseas, mission field for a period of six years before heading to South America. I felt strongly that we were supposed to take our seventh year off to have what missionaries call a furlough. I didn't want to. There was too much to do and I had a world to save. Since that time, I've been going downhill while accomplishing more than ever before at the same time. I feel like I am changing the world, while inwardly I'm continually slipping away from my conviction and integrity. Tanya is worried about my reputation and tries to cover for me, but I'm not making it easy for her. We have been firmly asked by our leadership not to return to our home in Micronesia. I want to go back and continue our work there but Tanya is refusing because she knows we need the inner-healing and counseling that we're being referred to. We just spent a week in Redding, CA at a marriage conference, then two weeks in Abilene, Texas for heart healing ministry, then another week in Seattle for Restoring the Foundations ministry. This was all recommended by our leaders at Iris

Global. It has been years since we have been in one place any longer than three months. I can't be settled again. I must keep moving. I am only one third of the way to my goal of visiting every country in the entire world. I still hide and drink, and hide that I'm drinking, whenever I have the chance. Our leaders are hesitantly allowing us to return to Mozambique to serve as staff/honorary leaders for the next Harvest School in Mozambique. So that is what we are preparing for now. My youngest sister, Grace, has been accepted to the school and she is going to travel the world with us and serve as our assistant. I'm so proud of her. She is raising support right now and has already purchased her flights.

He's turning the page again.

I don't know how this unfolds but I'm at rest in Him.

He's never not been good.

When I try to dictate my own course, I drive into a ditch.

He takes me into Him.

Where I see He's always been, all I need, in every time

I'm groaning deep within

He's stirring my spirit to desire more of Him.

I'm thankful for the pain that drives me to my knees crying out to Him.

Because I belong to Him.

CHAPTER 10

Africa Fail

Lost My Welcome
May 23rd, 2014

We're in South Africa now. I just met with our leaders over skype and they brought another couple into the call to introduce them as the primary leaders for the school. I can tell they don't think I'm ready and are trying to kindly replace us and remove us from the picture. I have manipulated my way this far and I don't want to give up my ground. We are the first ones here for the school and have already visited the new location to inspect the facilities since it won't be able to be held in Pemba this year.

Tanya and I came to stay at this ministry center in South Africa. Some of the staff for the school have already arrived here as well. A couple of them spotted me having a big glass of wine at a restaurant a couple miles from the center. I find excuses to get out when I can for good reasons, but I always have a hidden motive to drink and cover it up before coming back. I keep getting in loud arguments with Tanya at night. We stay in a tiny little hut surrounding by other huts with missionaries in them, and they can always overhear us. Word has gotten back to our top leaders and it seems that no one wants us at the school, because of me. I must find a way to make it work because my sister is coming. I cannot let her down. This is a dream come true for her and she has worked toward it for months now.

Alone in the World

June 10th, 2014

I just summitted Table Mountain, which is one of the New Seven Wonders of Nature. Cape town is my new favorite city in the world. I'm here alone, other than a number of other backpackers who stay at this hostel. Tanya and the kids flew back to the U.S. to attend her brother's wedding. She let me know they were not coming back after they arrived there. I was informed I was not welcomed to even attend the school anymore and my sister had to cancel her plans to come to Africa. I pretty much screwed everything up for myself, my family and most heartbreaking right now, my youngest sister. I don't know where to go or what to do, so I've been staying here at this hostel for almost two weeks now. I don't know how to get home. I don't know what home is. All I know anymore is the road. I am currently on a Jägermeister kick. I manage to drink nearly a gallon a day and stay normal. I hike the mountains and out-drink anyone I meet while doing it.

CAN'T STOP

June 23. 2014

We were given a nice mobile home to live in here in Jackson, Mississippi on O'Ferrell Street. I followed my leaders' instruction to come here for a time of restoration under this amazing spiritual mother and father we know from Mozambique. This is a famous neighborhood that has been transformed by the love of God through their service. I try to be involved but stay mostly distanced. It's scary to walk alone in these streets, especially at night. The neighborhood is still rampant with drugs. I am fearless and manage to find multiple excuses to go to the store and go for walks. I'm constantly drinking again and believing that I can hide it. I don't think anyone knows how to deal with me, so I just tend to isolate and hide. I'm sure that I am hurting my wife and children more than anyone else. I don't know what to do. I can't stop. I am broken and I don't want to admit it to anyone.

Vast Papa Divine

I sit here to behold you and quiet my mind. You're greater than greatness and stronger than strength. I will never fully understand You, but I will embark on this journey of discovery, diving into the adventure to experience the infinite realms of Your glory. As I see You, my first response is, "Wow"! You are always both speaking and listening at the same time. You think it's funny and You smile when I try to rhyme. When you smile, my heart feels warm. You speak in endless ways, however You want to, which makes it more of a mystery to converse with You. I hear You in the wind, in the grass, and in the trees... through my children, through the lost, and even through my enemies. You like surprises and so do I, which helps us get along together great. Is it possible to surprise you? Because I would like to surprise you and make you chuckle when you're sad. I can't imagine how it's possible to burn with every emotion simultaneously, but You do. Is it hard to be everywhere, all at once, helping everybody, forever? I can't imagine, but you exist outside the realm of our measurement of time... yet You live within and became like us to sympathize with our weakness and then overcame death. You are in the past, the future and the present all at once. Thank you for Your patience with me Papa. I love you Jesus.

CHAPTER 11

Sick and Separated

BROKEN
October 28, 2014

I spent the past five days in the ICU and have finally been transferred to a regular room. It's impossible to escape from here because I have tubes going into my lungs and heart. They say I have pneumonia, severe MRSA in my lungs, sepsis, and I've had two abscesses cut out of my chest. I have to carry a bag of liquid antibiotics in a fanny pack with a PICC line running into my heart for the next two months. I will come into dialysis once a week to have it changed out. I am okay. I can't wait to get out of here and be free again. Tanya separated from me and took the kids with her to Texas. I came to Washington because I didn't know where else to go. The doctor showed me an x-ray of my lungs and told me I have mild-Cirrhosis of my liver and that I need to stop drinking. They also let me know that my bone marrow shows signs of leukemia and needs further testing. I really want to be released so I'm trying to do these breathing exercises daily in order to be cleared sooner.

TRY AGAIN
December 25, 2014

It's Christmas! I've been doing much better and am recovering well. I started working again. It's awkward crawling under houses with this pack of antibiotics strapped to my waist, while it's connected and feeding medicine to my heart. I do it anyway because I believe I'm unstoppable and have to keep going. Tanya and the kids came back to

Washington for Christmas and we've reconciled our marriage. She believes the best in me. We are missionaries and are ready to get back out there. I've arranged for us to travel to Nepal after Christmas, where we will live and serve with our dear missionary friends in Kathmandu again.

Mercy unimaginably more than I merit

Makes me amazed by my marvelous Maker

Who Moves me in forward motion

Making the most of my mistakes

Marking my life with miracles

That turn the moments into milestones forever remembered

NEPAL
February 2, 2015

We rent a house in Kathmandu and I've been able to go on a couple treks into the mountains to minister in remote villages. There are missionary friends here who served with us in Europe, South America and Micronesia. We live separate from everyone and so I only join in the activities with the rest of the missionaries when convenient for me. Tanya is usually at home taking care of Zoe and Zion. There's a very cheap type of liquor sold everywhere in town. It's a clear liquid, so I put it in my water bottle even when I am trekking up mountains with the other missionaries. I don't think anyone knows. When Tanya gets upset at me and tells me I'm drinking, I just get angry at her for accusing me and I make a million reasons why she's wrong. She is right. I think I hide it well but it's hard to hide when you're 6'8".

STILL BROKEN
August 9, 2015

We only lasted a couple months in Nepal. It wasn't working out and we returned to Washington. I just came home from work and noticed suitcases packed in our apartment. When I asked Tanya about it, she told me that her and the kids have a flight to Hawaii tomorrow at 5am. My actions have hurt her and she believes that moving away is the best option for them right now. I've been very successful at my current work. God provided us a nice apartment here in town. I'm a foreman overseeing a large construction project and I get paid well. I work all day, come home, help make dinner and care for the kids, and then leave. Tanya and I don't really feel connected anymore. This is probably my fault for being so distant and continuing to drink. When I first got back here, I made a point to go to every AA or NA meeting I could find. Once I got busy with work, I would go after work every day. After a while, it became an excuse for me and I would chug a wine box before, in the middle of, and after every meeting. I would even share in the meetings and people loved when I spoke. I don't know if anyone knew I was still drinking. I wasn't lying about it. I just didn't tell them unless they asked. I found that the bar across the street from our apartment is much closer than the AA meetings, so I have made it my frequent stop instead. I don't socialize much and I don't drink to get drunk. I drink to forget, I drink to not feel, I drink to not think, and more than anything, I drink to survive. My body has developed such a dependence on alcohol that mere survival instinct causes me to keep alcohol constantly in my system or else I feel like I'm dying. I am upset at Tanya that she is moving away with the kids again. Though we have become disconnected and I do not feel in love with her anymore, I have always felt like it was worth it to stay together for the sake of the kids. Now I will have this apartment to myself. I'm not happy about this. I'm used to always taking care of people the best I can. I just learned that my dear missionary friend Tyren passed away from what was assumed to be Malaria. He is the missionary I lived with in Mozambique who caught me when I collapsed from Malaria and saved my life. I entrusted all the work I had done in Niassa to be under his supervision when we left. Now his wife and four children are still in Africa and no longer have him.

KAYLA
November 28, 2015

It's so hard being separated from my kids. I continue to manage this construction project and spend a lot of my time at the bar across the street after work. I'm still holding onto the Lord, as He never let's go of me. I started a Bible study in my apartment for a group of people from the bar. We open up to each other about our struggles and pray for one another. There's a particular young lady I love spending time with lately. I taught her how to play pool and she's very easy to talk to. I was able to talk about my feelings honestly and openly with her and I started weeping. It was the first time I've been able to cry in years. It's a back-and-forth struggle for me because I feel so much love for her but I'm still legally married to Tanya, so I often draw back. The bar had 86'd me and just allowed me to come back in, only I'm not allowed to be drunk or to be served any alcohol while I'm there. Kayla is trying to help me stop drinking altogether and I've been doing good for a short while now. She gave her heart to the Lord when I brought her to church with me and she is becoming more passionate for Jesus all the time.

Life's a difficult road with many obstacles along the way.

Choices to make, many paths to take.

We pass through dense fog at times and often it stays night for days.

Embrace each mountain top because another deep valley lies at the end of the decent.

In the deserts, we long for refreshing streams of cool waters

And in the lush gardens we often forget to be thankful

We may enjoy feasting in abundance while at the same time, in another part of the world, a mother watches her children starve to death.

There is beauty in suffering and hope of redemption remains a light in our hearts through the darkest hours.

Let's always rejoice in the provision and blessing from the Lord but never forget the millions drowning in sorrows and need, who are always reaching for a helping hand to stop for them and lift them up.

The joy of the Lord is my strength but my heart aches with deep groanings for my fellow men who are in the grips of the evil one. The peace of the Lord guards my heart and mind at all times, but I cannot settle until at least everyone has received an invitation to the wedding feast. In tragedy, God is always present to comfort all who reach out to Him. Let us never turn a blind eye to the poor, the orphan and the widow. We cross paths with people each day who put on a smile while inside they are hurting, possibly planning to end it all.

So God, let my heart be yielded to Yours and my spirit be open, that you can move through me to reach at least the least of these.

CHAPTER 12

Lost and Hurting

Kidnapped with Love

January 3, 2016

Currently, I'm in the backseat of my old discipleship ministry director's car. I was heading out for a side-job with my friend Brandon because I lost my job as a project foreman due to always being intoxicated. I was very surprised to suddenly see the director at my car window as Brandon and I were pulling away from my apartment. I hadn't seen him since I left his ministry years ago. He drove all the way from Kingman, Arizona because he was asked to kidnap/rescue me and take me back into his ministry for restoration. I didn't want to go, but he drove so far just for me, so I complied. We've been driving for a long time and I'm nervous. He has two hundred acres in the high desert hills where he is building a new ministry training center and wants me to help him. I am not well. I started drinking very heavily again and am frequently admitted into the hospital for safe detox. Kayla has completely cut me off now because I keep drinking when I promise to stop. I hurt her heart and I do stupid things when I'm blacked out. I say and do things I can only regret but I cannot change. My heart is experiencing more grief and turmoil than it ever has before because someone I love intensely is being taken out of this world and it's my fault. Everything is my fault. I have hurt everyone who loves me and I am being a horrible father to my children. I am a failure and I don't see any way to fix all I have broken. It would have been better if I never existed.

DEAD END

January 14, 2016

I'm now sitting in a San Francisco county jail cell while on a layover for my flight back to Washington. I lasted a week with at the ministry. They really tried, but I couldn't stay any longer. I walked off their land after I'd only been there for two days and continued 14 miles on foot before I reached a Highway. A homeless person let me stay in one of their tents in the desert for the night. The next day I invited them to drive back to the ministry to ask him if we could stay there. That was a mistake. They were turned away and I was left. I was told that if I wanted to leave again, I just needed to ask them and they would drive me to the bus depot. I asked them to drive me there the next day and they did. I didn't have any money, so I hitchhiked all the way to Bullhead City. I had no friends there, no money, and no charger for my dead phone. I tried to panhandle some change and was beaten to the ground by one of the members of the Hell's Angels. A young tweaker wanted to help me escape and told me there was a group there planning to kill me because of something I said. He took me to the middle of an open field where he said was the only safe place for me and we tried to sleep under a small tarp. I received word that my brother David had gotten a bus ticket for me to Las Vegas where I would be able to catch a flight. I walked to the bus depot at 4am and caught the bus. I slept under benches, outside of casinos in Las Vegas while I waited for my flight. My only layover was here in San Fransisco. I collapsed unconscious when I excited the ramp from the airplane and paramedics took me to the hospital. I guess I managed to drink too much again. The hospital released me soon after, as I convinced them I needed to catch my next flight. I couldn't find my ID after I left the hospital, so I stashed my bags behind a bush and went back to see if I had left it there. I didn't. It was just in a different pocket. When I went back outside, a sheriff the hospital had called was waiting for me and arrested me for public intoxication. I was not able to get all my belongings from where I stashed them and now I'm here in a jail cell trying to get released to go catch a rescheduled flight.

No Place for Me
November 11, 2016

This year has been a back-and-forth journey for me between Washington and Hawaii. I think I'm running from myself and I can't get away. Someone had moved me out of my apartment and all my belongings were gone when I first got back to Washington. Kayla took me into her home for a while but had to force me to leave again because I'm just a sloppy drunk now. I often try to stop drinking and fail. I carry too much that I need to forget. There is too much pain I don't want to feel. I don't want to exist anymore, but I can't just disappear. I thought being near my children in Hawaii would help, but I end up staying with the other homeless drunks here. I just want to be alone because I'm not good for anyone. Nobody wants me to keep drinking, but all I want to do is drink. I still trust God, but I don't know how to get out of this mess I've created.

My parents took me into their home the second time I returned to Washington. They had patience for me, believed in me, and did everything they knew to help me. I always had excuses to leave their house and do important things, but I just went to hide by the river and drink alone. I then covered my breath with gum and returned home to tell them all the important things I accomplished. I do attend a lot of AA meetings. I still always drink before and after each meeting. I often end up in the hospital because I have malt liquor for breakfast, lunch, and dinner and don't eat food for days. It finally catches up to me and my body crashes. The staff at both hospitals near my parents all know me now. They hook me up with IV's, tubes, and a heart monitor. I rip it all out myself as soon as I feel healthy enough and then I help myself out. My mom served me signed divorce papers from Tanya while I was there. I don't know if she really wanted a divorce but I am too much of a mess and I feel she made the right move. Kayla and I love each other and she has not given up on me yet. Everyone advises her to cut me off forever. Then she comes and finds me and gives me another chance.

I left Washington again after my dad picked me up from the hospital. He took me to their cabin and did everything in his power to prevent me from driving my truck again. Driving while drunk is dangerous and I'm always drunk. I was just released from the hospital here in Hawaii. I was still at a 0.55 BAC when I had woken up. This is fatal for most people but is now the second time my BAC has been at that level when waking up at a hospital. I normally stay at a 0.3 or higher just to stay normal. I never feel drunk anymore. My body has grown such a high tolerance to alcohol that I always need more just to stay alive and avoid withdrawals. It can be deadly to suddenly stop drinking when you have this level of dependency. I work a full-time construction job while in this condition. I brought a friend here from Washington to live with me on the island. He's very upset at me because I'm killing myself by drinking, while I drink to stay alive. The paramedics picked me up off a beach here last night because I wouldn't wake up and people thought I was dead.

ZURIEL
April 9th, 2016

Tanya recently gave birth to our third child. Her name is Zuriel. Now I have Zoe, Zion, and Zuriel. I went and held her earlier this morning. She was born at a mid-wife home-care facility in Leilani Estates. I regret that I wasn't able to be present for her birth but I am here now and Zuriel is very precious and adorable. I want to start making every effort I can to be more present for my children. It's more difficult now because I recently lost my car and do not have a steady place to live. Most people don't want me around anymore after they put up with me for a few days. I often sleep outside and wake up in strange places

My heart overflowing with expectancy

Leaping off the ledge of uncertainty

Into the realms of utter dependency

Upon my Father's faithful mercy

As I'm lifted by His wings upon the winds into His pure bliss ecstasy

His love that covers me and awakens me to His reality of eternity

I gaze into the awe of His majesty and feel my frailty

But He calls me royalty

Redeemed by His blood, He paid the cost at the cross

And from the grave He raised me up to be seated with Him in the heavenlies

Not by the labor of my striving and my trying to be holy

But by the response in my helplessness in faith of His grace

That fulfilled every requirement of the law, I am made new and left in awe.

CHAPTER 13

Kayla Tries Again

Receiving Help

August 10, 2016

I called Kayla during a very rough week of mental anguish, dealing with people taking advantage of me. This is normal for me because being in a constant state of intoxication makes me an easy target. The ones I love and try to help the most are almost always the same ones who betray me and take everything I have in the end. I don't know how to respond but to forgive them and love them again. Kayla's heart really went out to me, so she went against everyone's counsel and flew me back to Washington to help me again. Sometimes we play pool at the bar. We don't drink at all. I bring my Bible in with me and we play worship music through their stereo system. I've made it through detox and I'm beginning to feel good about myself again. Kayla serves at the church. She is a worship leader and is always digging into the study Bible I bought her last time I was here. I am very proud of her and she inspires me.

A Beam of Hope

June 6th, 2017

I started drinking again in Washington and all the doors began to close for me. I returned here to Hawaii to be close to my children. When I have nothing left, I am reminded that I always have them, and they are worth living for because they need their dad. I've brought my friend Brandon here from Washington to live with me. I have three small housing units on the ocean front. I sleep peacefully every night to the sound of the crashing waves outside my sliding glass door. Whenever I open my eyes, I stare at the moonlight glistening over the beautiful ocean as I lay in my bed. I was sleeping on the street in the village near where Tanya stays until a couple weeks ago. One day I became tired of being constantly sick from alcohol poisoning and sleeping with other homeless people so I got on a bus into town. I was hired at a sales job. I became the manager and now run the office. I guess that I'm a good hustler because now I have a home on the ocean and lease out a couple of the other units. Kayla is leaving her life in Washington behind and bringing her children, Tyrone and Autumn, to Hawaii. We are still in love and we both believe that we are supposed to be together.

Every day I hunger, a groaning longing deep within, panting for increase of intimacy. Wherever my eyes look, I'm searching to see You through Your creation and hear Your long song to me. Despising distraction that preys on every heart, I curse the spirits that seek to destroy me and fight with rest to receive my victory. Unattainable through my personal strength or discipline, but simply to believe the promise and reject the deception telling me I need to work for the reward that He already paid for through His work that I could never accomplish.

Wedding Celebration!
December 4, 2017

Kayla and I decided we want to spend the rest of our lives together. I proposed to her on a beautiful bridge at an ocean front park in Hawaii. We planned our wedding to be in Washington so that more of our family would be able to attend. There was only a short time to plan the ceremony and we had two weeks to prepare. My oldest brother, Joseph, officiated the ceremony for us. It's a miracle how everything came together so perfectly in such a short amount of time. Our family from both sides all came together in the most beautiful way to celebrate our life-long commitment to one another. Entering into a marriage covenant before God with Kayla has caused me to experience God's redemptive nature in my life in a deeper way.

We had packed up and secured our belongings before leaving the island. After we left, we received news that my pickup was stolen and our home was taken over by other people. Hawaii is home to us and we had no idea where we were going to stay when we returned. This was a troubling obstacle to be facing while attempting to stay focused on our wedding. Our wedding was two days ago and our return flight to Hawaii departs in two days. Today I was contacted by the owners of an apartment complex on the island and we were asked to become the managers. It's an apartment complex for low-income families and college students. We accepted the job because it was an answer to our prayers and we will have a place to live. I think Kayla will do very well with the challenge of managing this new place. We are married now and setting off for a new adventure and a long-term honeymoon in Hawaii!

Struggling with Compromise
March 30, 2018

It's rarely peaceful while living in the center of this apartment building over the last months. Kayla is an extremely capable manager and does well at her job. People are knocking on our door every day and throughout the night with various needs. I often help calm domestic disputes, assist with evictions, and interpersonal community drama. Most days, the Hawaii police department show up in large number to respond to disturbances. I've renovated beneath the building and set up a church. Kayla is pregnant and God is good! I've been drinking only light beers on a regular basis. I like Bud Light Lime. I decided that if I stick to light beers, I stay a lot better. It helps settle my stomach ulcers, calms my acid indigestion, and it's nice to drink cold beer on a hot day in Hawaii. Sometimes I still fudge and drink little one-shot alcohol bottles and wine coolers in between my beers. The owners of this apartment building live on the mainland. Kayla and I are the face of this entire place and it's always a challenge. There is often drug trafficking going on under our noses and I'm still learning how to respond to it.

A Lily Blooms
June 30, 2018

Lily Grace has been born!. Kayla did so amazing while giving birth to her. I was at her side, holding her hand, and am now the proud father of yet another beautiful daughter. I spoke with Kayla about feeling like we need to leave the island. She immediately confirmed with me that she was hearing the same thing from God. This whole side of the island has become chaos after the volcano erupted again much closer to us. It's not safe to walk down the street anymore. Thousands of displaced people have flooded into our town. The air is thick with smoke and it's hard to breathe. I managed to overcome my addiction to Vicodin that

I had been generously prescribed due to my serious back injury. My dependence on them steadily grew heavier, and mixed with alcohol, they were changing me into someone I am not. I am supposed to be the leader. I often disappear for days at a time on drinking binges. When I drink alcohol, mixed with my prescription drugs, it causes me to become psychotic. I've been diagnosed with PTSD, severe anxiety, and depression which I've been prescribed many types of drugs for. I want to stop them all. More than anything, I need to stop drinking again. We are preparing to quietly leave as soon as Lily turns seven days old and is legal to fly. Things are getting bad here in many ways. I'm sure it would be much better if I weren't drinking so much, but I don't allow anyone to keep me accountable. I just do what I want and everyone is getting hurt in my wake. I don't know how to ask people for help. I like to handle everything myself, but I'm not handling it well.

Searching for a Home
August 27th, 2018

Kayla and I are trying to adjust back to life here in the PNW. We left Hawaii as planned when Lily turned one week old. We don't have our own home here yet. We're starting over from scratch again. We first stayed as guests at my friend's apartment. I started searching for work while still binge drinking off and on. It was hard to feel settled without a place of our own. My parents recently took us into their home again because that apartment was an unhealthy situation for our family. Kayla and the kids moved into their house first, but I was not allowed because I was usually intoxicated. I walked the streets for days, feeling lost. Both my feet had swollen to the size of footballs. They became infected and were bleeding until I could walk no further. After I was released from the hospital on crutches, my mom decided that maybe I had gotten the point that I can't keep running any longer. I made a commitment to them to have zero alcohol and they are helping us get back on our feet again. We've been here a couple of months now and

I have been able to avoid alcohol completely.

God, I pray and ask you that if I start drinking again, You will knock me over the head with a 2x4 so that maybe I'll get the point.

2x4 TO THE HEAD

September 3rd, 2018

I became so angry last night! I was angry at my wife because she took our kids to a birthday party and called to let me know they would not be coming home because she locked her keys in the car. I could tell all the adults there were partying, so it was probably better that she had locked her keys in the car. Nevertheless, I became very upset and walked to where she was at to try and help get the keys out and bring them home. On the way there, I bought a couple wine boxes and chugged them down. I was unsuccessful at getting the keys and had a conflict with the people from the house they were at. I left on foot again and woke hours later laying on the sidewalk. I have a severe concussion and bruises across my head. I don't know what happened, but I am pretty sure I was knocked over my head with a 2x4. I will recover and I'm sure I'll be fine, but inwardly I feel a lot of turmoil right now.

My son, do not make light of the Lord's discipline, and do not lose heart when He rebukes you, because the Lord disciplines the one He loves, and He chastens everyone He accepts as His son. Hebrews 12:5-6

CHAPTER 14

A Year of Living Hell

Lord I pray that this coming year, above all, you would help me to be a good husband, father, and son. I know that you still have more purpose for my life, and if I cannot get it right with my family then how will I continue forward to do the other things you've called me to? I need you more than anything Lord. Please help me to stay sober and focused.

Stiff-Necked and Tormented

December 28th, 2018

I am so lost right now. I'm unhappy because Kayla and my kids are at my parents' house but I'm not welcome there with them. I've been binge drinking often now and I only stay sober for a few days after I get out of the hospital. I think everyone has given up on me, and I've given up on myself. I've admitted myself into a number of recovery homes, only to leave a few days later so that I can drink again. I don't know how to manage my life. I want to take care of my wife and children, but I am doing a horrible job at it. I am failing completely and I just want to disappear. A crowd of black crows follow me wherever I go. I'm very annoyed as they are non-stop cawing at me. There are shadow people following me everywhere. Certain people can see them, while most cannot. I know why they're not leaving me alone. My eyes are open to see everything in the realm of the second heavens and I drink more to try and ignore it. I've been passing back and forth from death to life and there are different locations I'm brought to by these shadow people. It is tormenting and deceptive in each place because it seems that it is impossible to leave once I'm there. There are many other lost souls in these realms who try to urge me to let go and stay there where they are now stuck. In these battles, I hear my dad's voice praying for me from above and something about this is what pulls me back into my body. I don't know what it's going to take for me to finally get it together. I want to stop drinking but I can't. I feel like I will die without alcohol, as it is killing me at the same time. I guess that being hit over the head by a 2x4 wasn't enough to teach me the lesson I needed.

When I was a young boy, I used to have horrifying dreams of being stuck in small places. These dreams caused me to have claustrophobia until today. People who love me keep trying to help me by causing me to be forced into mandatory, involuntary treatment. Each time I manage to escape or manipulate the system to get myself out as soon as possible. I hate confinement. This is why I feel most at home when I am staring at the night sky or on the coastline looking at the horizon over the ocean. I'm supposed to be with my family. I'm angry at my parents for forbidding me to come to their home and I keep going back to try and get in. My wife and kids are there and we are supposed to be together.

A BIGGER 2x4

February 1, 2019

They won't let me leave this hospital. I have tried a few times but the hospital placed 24/7 surveillance and security over me. I can leave my room to walk around this floor of the building as long as the guard walks with me. I made it outside to the street once in just my hospital gown and then was escorted back in by a group of security guards and nurses. I got hit over the head with a bigger 2x4 this time. I think God is trying to get a point across to me. I woke up here in the ICU with Kayla at my side. I plucked the metal staples out of my head as people tried to explain what had happened. I was hit by a pickup while trying to cross the street on foot. I had attempted to go into my parent's house to be with my wife and kids and the police came. They told me I had to leave and so I did. This accident happened a quarter mile away. I was on life support for the first 18 hours and in the ICU for the following five days. I think I'm ok to be checked out now but the doctor will not clear me yet. I am on all sorts of meds here and my head isn't right. Well, I guess that my head was literally split open just recently. I am very lucky to still be alive. Kayla often comes and spends time with me here. We play cards and I have not regained sanity yet. I try to pretend I have. A lot of people come to visit me here.

Incarcerated

June 4, 2019

I don't know why I'm sitting in the country jail again. This is the fourth time I've been back so far this year. I can't stand it here because I hate confinement. I would escape if I could. I'm not trying to be a criminal and I don't feel like I belong in this place. I was praised by some of my cellmates because It took twelve officers to restrain me as soon as they took my handcuffs off in booking. They had to call for backup three times, and that was inside the jail walls. I felt a little puffed up about it in the moment, but I never even remember what happened before I wake up in a cell. Even then, I don't remember. I just read the police reports and hear stories. Since I had the traumatic brain injury in January, when I drink, I immediately black out and then I keep going without knowing who I am or what I'm doing. I can't drink like I used to anymore. I finish a gallon of whiskey a day when I'm free. I try to reconnect with my wife and kids, but no one wants to be around me when I'm drinking. Everyone just tries to save me. I think they are trying to save me from myself. I am my own worst enemy. I have court tomorrow and I'm hoping the judge releases me again. I don't know what I'm going to do. I just want to get out and be free again. I don't want to go into a program. I can do this myself. I'm a capable person and I always get back on my feet.

Birthday Wishes
August 2, 2019

I am here AGAIN in county jail and today is my birthday. The best part of today was that my wife answered my phone call and she and my children wished me a happy birthday. I continue to fail them. She has filed for divorce already. She is eager to not be tied to me anymore. She probably wished she would have listened to everyone else in her life when they urged her not to NEVER give me another chance. They are out there just trying to get by while I am in here because of my own stupidity, and I am not able to help them at all. Happy Birthday to me! I have court in a few days. I have high hopes to get out again.

Excruciatingly Painful Disappointment
August 7, 2019

Well, I got out and went to see Kayla and my kids. Kayla is managing a women's recovery home. I learned that she is back in a relationship with a man who is helping her run the office at this recovery ministry. I've spent all my energy trying to prepare to make this last time, the last time drinking and in jail. I developed a good plan of how to stay sober and provide for my family. I left her and the kids after visiting for a bit and receiving this news. I am throwing everything back out the window. I can't take this and I feel like everything that was coming alive inside of me just died again. I'm done. I picked up a backpack full of whiskey and got another hotel room. I don't want to ever see anybody ever again, forever. I don't blame her. I have been gone all year, drunk or in jail. She has been left alone. I was called out of my cell for a phone meeting with her about child support. Our divorce will probably be finalized soon. It's not what I want, but I'm not in control. I can't even control myself. I prayed at the beginning of this year for God to help me become a good husband, father and son. I've been stripped to the core in those exact areas and I am the worst husband, father, and son ever. There is no hope for me.

Revolving Doors
August 15th, 2019

It only took two days this time and I'm back behind bars. It was supposed to be a catch and release for a minor offense. I offended the sheriff, so I was made to wait in custody until I had a court appearance. The judge knows me now and gave me the maximum penalty possible. I was supposed to be released the same day but I've been sentenced to 90 days in custody instead. There is a death warrant out for me in this jail because I offended the wrong inmates, so the jail staff are working to move me to the Work-Center location quickly. I have written a letter to the judge and will continue to write him letters asking for mercy until I'm released again. I have lost everything, again, and I know it's all my fault. I can't get it right. I don't know why I didn't die, or just stay dead, a long time ago. Through it all, I have a trace of hope left. This trace of hope is all that keeps me holding on. I have six beautiful children.

Death has come to visit me here and keeps offering me a deal. This spirit can present itself seductively enticing, while it pulls on me through a lust for power and the freedom to unleash all the rage and hate that burns in me toward all those I've blamed for my pain. Under its dark shadow, these actions of rage are presented as justifiable and necessary. This is the same spirit who Jesus eternally defeated when He spent three days in the grave. This battle has already been won and now everyone must choose sides. I'd rather be on the team of the Great Champion. During these dark battles and encounters in the 2nd heaven realm, God is constantly present and reminds me of Truth. To take the deal and align myself with Death in order to fulfill the vengeance the dark side of me burns for, would cause devastation, creating a wake of eternal regrets in the sea of life around me. I did not sell my soul in this deal I was offered and instead I let that part of me die. In its place, God gave me a new level of His Agape love. Through the darkest valleys, while suffering the deepest wounds, I have become closer to the Father than I could have imagined possible before

A river runs deep in the spirit of the redeemed.

Living water flows continually from the heart that's given to the King.

Along the banks are trees with leaves that bring healing to the nations.

A divine journey of discovery into the Almighty searches out infinite realms of glory

Never settle down into the land of earthly comfort and security.

You must continue this pilgrimage if your quest is for your destiny in the land flowing

with milk and honey.

Your own strength will fail, and there- grace is made perfect in your weakness that will

cause your arm to lift until you have carried your blazing torch across the finish line.

To a heart that embraces the deserts of life, The Divine One reveals Himself through a

blazing bush and causes water to gush forth from stone.

There in your wilderness the heart is exposed, hopefully revealing your readiness to

receive His Kingdom promise in fullness.

Steadfast and faithful, endure the stripping until in nakedness, shameless, you are

caught into the paradise of blissful ecstasy, intimacy between you and Almighty.

CHAPTER 15

-But God-

Refiners Fire Testimony, by Kayla Gellatly

I remember seeing him around, but his face was normally hidden in the shadows of a corner in the dimly lit bar. He didn't talk much; he certainly wasn't like the other guys here, who were mostly looking for someone drunk to take home. I was so intrigued by this man! He would never speak a bad word about anyone, so kind and gentle, loved unconditionally, extremely generous towards everyone- including people who slandered him, or stole from him. I thought he was crazy!! Little did I know that I would soon fall madly in love this giant, mysterious, adventure seeker. I also didn't know at the time that even at 23 years old, I had no clue what real love was, because I didn't know the One who *is* love. There was something unique and different about Jesse, so I wanted to get to know him better. We spent many hours together as he taught me how to shoot pool, and well! He was able to open up and talk to me, in a way which he described never being able to communicate to anyone in his life.

I decided to go to church with him since this relationship he had with God that couldn't be hidden was unlike anything I'd ever witnessed! Miracles and favor surrounded Jesse, even in his darkest of days. I was astonished to watch as things that seemed so impossible to my worldly mind, were the everyday normal, and somehow I just missed it for my entire life. I remember being extremely sick, couldn't lay flat, and could hardly doze off before being jolted awake by horrible coughing fits that wouldn't let up. Around 5am, Jesse rested his hands over each of my ears and said a simple prayer. After his prayer, he left my bed and I was able to lie down, FLAT, and slept peacefully and soundly until around 8am when I awoke to my children ready to start the day. I was astonished! This was the first miraculous healing that I experienced, and it made me realize that this was real. God is real.

Miracles are real. It's not for the times of the past, and it's not just special occasions.. this is everyday life when you know who you are and you walk in the authority granted to you by The Almighty Papa God.

Although there was so much excitement at the beginning of our love story, it didn't come without also great hardship. We both loved whiskey, a bit too much. I loved Jesse so immensely but I was continually hurt by the aftermath of the binge-drinking that he had turned to in order to forget everything that troubled his mind the most. He wanted nothing more than to be close to his kids, but it seemed so difficult since they had moved so far from him a few times already by this point. We had both quickly gotten to the point of drinking every night of the week, and I noticed that this was spiraling down fast. I had decided to cut back, and I would only drink on Friday and Saturday nights, and would drink water or energy drinks the rest of the week. This came along with wanting Jesse to quit drinking as his worsening habit usually led to sloppy behavior and would soon be very difficult to handle Jesse in his intoxicated state of being. I was so torn day by day. We had mountain-top highs and deep valley lows. I've now come to realization that this may very well be a pattern in this life journey, especially when two extremists choose to do life together. There had become a point however, when I decided that no matter how much I loved him, I could no longer bear the tremendous amount of energy it took to deal with this seemingly never sober Jesse. I knew I could love him from a distance, and I knew he needed to be with his children, and so I tried to call it quits and told myself I would never let myself get into this situation again.

That was the plan; however I couldn't remain in that place of hurt forever. I was learning how to forgive, and learning about this unconditional love that was absolutely foreign to me up until this point in my life. Our life together has certainly not been "easy", but it's well worth it. An entire book could be written about the miraculous things God has done in our lives in just the last 6 years, and the beauty that comes from God's faithfulness even when we are all so messy. God is the one who can take something broken and turn it into something beautiful. I am so grateful to have met a man who loves God with his entire being and yearns so much to represent Him well, and does. I know I'm not perfect, and there were times that I had given up on Jesse, and truly didn't ever see us working out. I was devastated that I had married a man that I thought had gotten alcohol under control because he was not drinking liquor anymore, just to find that we hadn't seen the worst yet. He sobered up well for a few months, but kind of started drinking a little again. After he was hit by a vehicle when crossing the road, I wasn't sure he would live through it. I lied next to his bed nearly the entire 5 days he was in the ICU. He suffered a traumatic brain injury, fractured skull, lacerations to his head, bleeding in his brain, internal bleeding around his liver, bruised ribs, along with plenty of road rash. His 3rd day in the hospital happened to be my birthday, and he had still been completely unresponsive up until this day. At one point in the evening, tears filled my eyes and were running down my face. I leaned in to kiss his forehead and said, "I love you", and in just the faintest mutter I heard him trying to say it back! He really does seem to spoil me on my birthday, but I don't know if this one can be topped! This gave me the slightest bit of hope that he would recover and walk and talk again someday.

By the grace of God he recovered well, and very quickly. However the lasting effects of the brain injury didn't go away as fast, and he often ended up even more of a drunken mess than ever before. I really didn't know how it could have gotten any worse, yet here we were. The prescription drugs to "help" because of the recent accident, paired with alcohol, were a recipe for disaster. This time led to him being in and out of jail, and after about 5 different times back in county, after making a plan, saying whatever he needed to, to just get out and then go drink again and do it all over, I was fed up and done. I had 3 children to now raise on my own. I was managing a women's recovery home, and *that* is enough of a challenge as it is. I couldn't allow myself to be hurt by Jesse's self-destructive path any longer. It came to a point where I lost faith in us and I filed for a divorce. Anytime I saw Jesse anymore, he made me so mad I wanted to physically hurt him. This I know, is not okay, or healthy for any relationship, or for our kids to see. I couldn't put them through this any longer. I didn't know who he was anymore. The man I once saw in there.. I couldn't find him anymore. I didn't know if he still existed inside of this person who looked and sounded like the man I married.

I counted down the days. I couldn't wait for the divorce to be finalized. I felt defeated. I was tired of stressing out over what he was doing or where he was going for the short amount of time he spent out of jail, because I knew he could stop at any convenience store and buy a box of wine or shots of liquor, and then lie extensively about it even though it couldn't be more obvious. I was ready to put all that behind me and work on building myself back up into the strong woman I once felt I was.

But GOD!

I had still kept coming to our little church this entire year, just me and the kids. I didn't always feel like it, I was pretty mad at God. I wasn't feeling heard but He reminded me one Sunday that Jesse wasn't dead yet, and so it wasn't too late for a change… I literally said to Him, well that sounds crazy, I've tried this countless times; but my way isn't working! Alright…

I can't recall a single person who thought it was good for me to jump right back into trying to work out my marriage without some sort of time, program, or some type of proof that he was going to do something different this time. But for the first time in a long time I heard from the Lord so clearly and had to follow what He said and not the knowledge of the world. Jesse was still incarcerated this day, and when I answered his call a little later and told him I wanted to give our marriage one last shot he sounded both excited and unsure as to whether that's what I really said. He asked again to clarify. He called again later that afternoon to ensure that this was real. It was November 2019, the divorce had been filed for over 3 months already, and he had been telling me, even when I was completely bitter toward him, that God had promised him a home for us THAT year, and that even if we separated he would make sure I had a home. Seemed pretty impossible to me since I wasn't planning on being involved with him and couldn't see a possibility with the circumstances given. But God is forever faithful! On December 31, God provided the finances needed for a home, and then on January 11, just a week and a half later we would be signing the lease to our new house- one that was everything we had been praying for and more. By nothing short of a miracle, our lives were turned around that one day that I heard God. Jesse was a new man, a man who I would soon come to find was even more perfect for me than I knew – I hadn't ever known Jesse sober. The last two years I've been able to get to know this husband of mine who is restored through the power of God, and our family is renewed, being stronger and closer than ever before.

Now more than ever, I can see God working even in the dark times, even when we don't feel Him or hear Him. He remains faithful, and He continues to refine us, sometimes in areas we didn't even know needed work. God does not forsake you, and He will never forget about you. Keep looking up and keep pressing on. The night gets darkest before the dawn.

CHAPTER 16

A New Heart, A New Dawn

Dawn of a New Day
January 20th, 2020

Released from behind locked doors and free at last! I have stayed sober and focused for over two months now. I will do whatever it takes to never go back to jail. I told myself that all seven times I was released from jail last year. The judge finally responded to my letters and released me in November, eleven days early. I got a hotel room and was nearly out of money. I didn't feel like I was off to a good start. I just didn't know what to do with myself. I've had frequent long conversations with God over the past year. I shared with him how I believed the one thing that would really help me become more stable and focused would be to have my own home. I purposed to still get a house for Kayla and my kids before getting one for myself. Our divorce was not finalized yet. Even though we were not staying together, I owed it to her to get her a home and I planned to because I said that I would. She has suffered so much because of me. I trust God and my faith in Him is still strong, for He remains faithful, despite my frequent failures.

I never gave up on my marriage and prayed earnestly and frequently for Kayla and for our relationship to be restored. I was dreaming with God and made a list of what I desired in our new home. It needed to be a house and not an apartment, so that we would have our own backyard. It needed to have three bedrooms so that Tyrone would have his own room. I wanted a garage, a fireplace, an outdoor firepit, and the list went on. In November of last year, Kayla informed me that she wanted to try again at our life together. We had nothing to start with except for God and each other. Our divorce was rescinded by the end of 2019. We recently moved into our new house in a neighborhood called Sunset and the sun is rising on the dawn of a new day in my life. God provided us a home with everything from my list including an outdoor firepit. We now have everything we need and our children are very happy in our new home. By working through the most difficult times and not giving up, we are now stronger together than ever before. God has literally restored everything that I had lost. All that was broken, dismembered and destroyed by my failures is now healed and stronger than it was before I broke it.

Here we go again.

Leaping off the edge of this cliff, not knowing where it will end,

But in Your arms again.

I didn't know I had these wings until I jumped to what could have been my end,

And now I soar to higher heights than I have ever been.

No More Running
July 10, 2020

Over the past couple years I've shared with Kayla my desire to take her and the kids to see the Grand Canyon. Now we are here. We drove from our home in Washington and were able to stop and visit family in different states along the way. I don't drink anymore, not even a drop. I also acknowledged the damaging effects all my prescription drugs were having on my mind and I stopped taking all of them. I've found my mind, body, and soul to be increasingly healthier without them. My brain, body, organs, family, mind, and life have all been completely healed and I don't hide anymore. When I carried the shame of my failures, it caused me to hide from others. The best way that I could find to hide was by drinking alone to forget everything until I passed out. Jesus reminded that He bore all my shame when He hung on the cross for me and that there is no reason for me to bare it again. It no longer exists. I have nothing to hide from. I have messed up and made countless mistakes, and I lay them out for everyone to see. I have nothing to run from. I am forgiven, washed, and clean. Staring over the Grand Canyon at sunrise with my family is causing me to remember these things. I hear the Lord whisper to me, "Son, I never let go of you. I never gave up on you. I was always right here waiting for you to look up again".

I realize that I had been blaming everyone else for my problems. This caused me to carry resentment and I stopped liking people in general. I could never trust anyone again because I thought everyone was out to control me and take advantage of me. This caused me to isolate myself into a hole of self-pity. When I looked inwardly, I realized that I should blame myself for my failure because it was really all my fault. I was selfish. I hurt everyone around me. I fail at everything. I am not a good person. Shame and blame merged as a large shadow hovering over me. Then the voice of Jesus called to me, "Jesse!". I looked up and saw a doorway open with light shining out into the darkness all around me. His voice spoke again and said to me, "We are not done yet. We have a journey ahead and you have much more that I want you to do. Come, I am going with you." I got up, brushed myself off and started to follow Him. Blame belongs with the shame that I no longer carry. I am reminded that Christ not only bore the weight of my shame, but also my blame. I am not to carry it anymore. I do not blame others. I cannot blame myself. It is an offense to God when I try to carry what He already suffered and died for to deliver me from. I failed, but I am not a failure. The price has been paid for the things I've done wrong and I am pardoned. I don't deserve this grace, but Jesus bore what I deserved so that I may have what He deserves. I am a victorious warrior and an overcomer.

Climbing out of this dark hole, my eyes struggle to adjust to the light that has been too long avoided. As I pry the coils of shame and blame out of the flesh of my shoulders, my heart begins to shake and break apart the hardened clay covered surface that has trapped and hidden the life inside. Breathing fresh air again begins to flood my inner-being with a hope that moves to exterminate the infestation of maggots and roaches that made this dirty, dark place their home.

Malnourished and aching from lying trapped inside that filthy hole, I push myself to crawl as far away from there as possible. The clouds part and the sun shines forth, illuminating a straight yet narrow path before me. Rising to my feet I begin, one foot in front of the other, one step at a time, determined to not look back at the regret behind and find what lies ahead, sure that anything is better than the darkness that had left me for dead.

I hear the soothing sound of moving waters and I come upon a cool stream. My throat is dry and beyond thirsty. I drop to my knees and drink deeply. Immediately I begin to see and think more clearly. New strength and energy fill my body and soul. I think to myself, "How did I end up in that hole? I knew way better, I should have been stronger, I must be stupid, I wasted so much time, I failed and am a disappointment." A strong yet comforting voice from across the stream commands, "STOP". As I look up, I see an old friend, my Big Brother, the one I was too ashamed to be around because I thought I let Him down. He smiles and shakes His head. "That is behind you," He says. "You are not to go that way anymore. There is no need to look there because we are going a new way now. We have a long way to go and so much more to do. I have been waiting for you and have a meal prepared. We are going together and you are stronger now than ever before". As He took my hand, my whole being flooded with warmth and light and we began to walk with resolute purpose and joy.

Proverbs 24:16
For though the righteous fall seven times, they rise again.

Life Lessons Learned
December 28, 2020

There has been a lot of chaos going on in the world this year, but I have perfect peace. I've had an amazing year and my vision is clearer now than it's ever been. I haven't drank in over a year. To others, it appeared that this was my major problem and was why my life was in such shambles. There is truth to that, but alcoholism was merely a symptom of more serious, deeper issues I was trying to cover up and hide from. I was self-medicating severe PTSD and the pain that I had on every level. I have learned to just face the pain and deal with it instead of trying to hide. If I wanted to drink again then I would just go drink because I'm stubborn like that. I drank until I finally had enough and I don't want it anymore. If everyone around me were drinking and having fun, I would not even be triggered. The desire has completely left me and I know myself better than to open that door even a little. I am sober today because of a simple decision fueled by desire, to just stop. More than this, it is the grace of God. I asked him to take the desire for alcohol away, and He did. This story is not about alcoholism, addiction, or failure. My story is called *The Journey Unveiled* because the veil of hiding is lifted. God set me free, to be vulnerable enough to walk in transparency. My story illustrates the faithfulness and mercy of God despite my frequent failures. No matter how far you run or how deep you sink, you can rise again and be restored. The Lord is full of mercy and compassion and pleads for us to come home. Leave all your baggage behind, arise and run to Him. No matter how many times you fail, get back up and keep moving forward. I am not alive today because of one miracle. I am alive today because of a multitude of endless miracles. I can run from my problems but I cannot outrun the love of my Father God. I crave adventure and I love my life. I now spend most of my time at home, praying and taking care of my family.

My highest purpose is to invest all I am into my six amazing children. Zoe, Tyrone, Zion, Autumn, Zuriel, and Lily are young world changing diamonds. They are my legacy and God has resurrected me and kept me alive for them.

A NEW HEART

October 15th, 2021

For seven years I've carried a dream to visit the Middle East. It is the one region of the world I had never been to. I am here now and this dream has been fulfilled. Jordan is the 57th country of the world I've visited so far. I've now driven through the entire country, walked the Bible and experienced first-hand many of the locations where the stories took place. Daniel is our guide and he asked our team to each share what we desired to receive most from this journey. My response was that I desired to have a new encounter with God that would leave me forever changed.

Everything about our journey through Jordan has been awesome. God was revealing Himself to us in powerful ways each day. I wasn't sure I had the encounter I was hoping for, until today, our final day on the journey. We went to Peniel, to the very spot where Jacob wrestled with God all night and was renamed Israel. Jacob told the Angel of the Lord that he would not let Him go until He blesses him. Jacob wrestled with God and prevailed. At daybreak, he limped away much slower than he was able to move before God blessed him. Peniel means **Face of God**. Daniel encouraged us as we separated to pray at Peniel, "The blessing God gives you may come very different than you expect or desire".

Genesis 32:22-31

[22] *And he arose that night and took his two wives, his two female servants, and his eleven sons, and crossed over the ford of Jabbok.* [23] *He took them, sent them over the brook, and sent over what he had.* [24] *Then Jacob was left alone; and a Man wrestled with him until the breaking of day.* [25] *Now when He saw that He did not prevail against him, He [c]touched the socket of his hip; and the socket of Jacob's hip was out of joint as He wrestled with him.* [26] *And He said, "Let Me go, for the day breaks."*
But he said, "I will not let You go unless You bless me!"
[27] *So He said to him, "What is your name?"*
He said, "Jacob."
[28] *And He said, "Your name shall no longer be called Jacob, but Israel; for you have struggled with God and with men, and have prevailed."*
[29] *Then Jacob asked, saying, "Tell me Your name, I pray."*
And He said, "Why is it that you ask about My name?" And He blessed him there.

[30] So Jacob called the name of the place Peniel: "For I have seen God face to face, and my life is preserved." [31] Just as he crossed over Penuel the sun rose on him, and he limped on his hip.

I went into the middle of the river which was knee high. I struggled upstream and downstream as I absorbed Peniel's atmosphere in. I came out of the water, hid behind a large bolder, and cried out for God to meet with me. I didn't know what to say, but I know I needed Him more than ever before in my life. He met me there and here is the description of our interaction.

"Father, PLEASE HELP ME! I am here but I don't feel. I don't want to just take up space. I want my life to matter and make a difference in this world." Suddenly, I was presented with the opportunity to have a heart again. I consistently feel strongly what is happening in the spirit realm, but I have not felt real emotion in a long time. I remembered how my wife told me I seem always guarded, like I keep walls of protection around my heart. I know that I was not putting up walls and that I literally had no feelings left to protect. My heart had been burnt to a complete crisp and the ashes were blown away. I have not felt true happiness, compassion, or cried for years.

I felt like I was standing face to face with God and I knew I had to ask Him for a new heart if I wanted one. When poised with that question, I was suddenly aware that I wasn't sure if I wanted it. I haven't felt for a long time and having a heart again seemed like a giant risk that I needed to consider the cost of before asking. Pain hurts! And emotional pain and wounds to the heart hurt much worse than a broken leg. I got stuck here for a few minutes behind the rock, going back and forth with God before making the decision. Not feeling emotion is easier than feeling. If I have a heart again, It is likely to be bruised and beaten by everyone as my previous one was. Father God had a heart for me in His hands and He offered to help me care for it if I would trust Him to put it in me. Reluctantly at first, I asked Him for a new heart. I prayed and asked that it would beat with His heartbeat and yearn for what His heart yearns for. He placed it inside of me, not as His heart, but as my own new heart, and I committed it to Him. His heart longs for relationship with all mankind, breaks over the injustices in the world, and rejoices from the affections of His children. If I am to love Him and to love mankind with Him then I want to feel His joy and His pain.

I was not expecting this before I visited Peniel. Everything is going to be new and different again. I had the encounter with God I hoped for from the beginning of the journey and I sense fresh, new adventure ahead.

God, my heart is yours. If it gets crushed again, please let it be crushed for you. If I die, let me die for You and while I live, I will live for You.

ABOUT THE AUTHOR

This book, The Journey Unveiled, tells you about the author. If you would like to learn and read more from Jesse Gellatly, please visit the authors website at JesseGellatly.com.

Made in the USA
Middletown, DE
03 February 2022